STOCHASTIC MUSINGS: PERSPECTIVES FROM THE PIONEERS OF THE LATE 20th CENTURY

STOCHASTIC MUSINGS: PERSPECTIVES FROM THE PIONEERS OF THE LATE 20th CENTURY

Edited by

John Panaretos
Athens University of Economics and Business

*(A Volume in Celebration of the 13 Years
of the Department of Statistics of the Athens University
of Economics & Business in Honor of Professors
C. Kevork & P. Tzortzopoulos)*

LAWRENCE ERLBAUM ASSOCIATES, PUBLISHERS
2003   Mahwah, New Jersey   London

Camera ready copy for this book was provided by the author.

Copyright © 2003 by Lawrence Erlbaum Associates, Inc.

All rights reserved. No part of this book may be reproduced in any form, by photostat, microform, retrieval system, or any other means, without prior written permission of the publisher.

Lawrence Erlbaum Associates, Inc., Publishers
10 Industrial Avenue
Mahwah, New Jersey 07430

Cover design by Kathryn Houghtaling Lacey

Library of Congress Cataloging-in-Publication Data

Stochastic musings : perspectives from the pioneers of the late 20th century : a volume in celebration of the 13 years of the Department of Statistics of the Athens University of Economics & Business in honour of Professors C. Kevork & P. Tzortzopoulos / [compiled by] John Panaretos.

p. cm.

Includes bibliographical references and index.
ISBN 0-8058-4614-X (cloth : alk. paper)
1. Statistics. I. Kevork, Konst. El., 1928– . II. Tzortzopoulos, P. Th.
III. Panaretos, John.

QA276.16 .S849 2003
310—dc21

2002040845
CIP

Books published by Lawrence Erlbaum Associates are printed on acid-free paper, and their bindings are chosen for strength and durability.

Printed in the United States of America
10 9 8 7 6 5 4 3 2 1

Contents

	List of Contributors	vii
	Preface	x
1.	Vic Barnett: *Sample Ordering for Effective Statistical Inference with Particular Reference to Environmental Issues*	1
2.	David Bartholomew: *A Unified Statistical Approach to Some Measurement Problems in the Social Sciences*	13
3.	David R., Cox: *Some Remarks on Statistical Aspects of Econometrics*	20
4.	Bradley Efron: *The Statistical Century*	29
5.	David Freedman: *From Association to Causation: Some Remarks on the History of Statistics*	45
6.	Joe Gani: *Scanning a Lattice for a Particular Pattern*	72
7.	Dimitris Karlis & Evdokia Xekalaki: *Mixtures Everywhere*	78
8.	Leslie Kish: *New Paradigms (Models) for Probability Sampling*	96
9.	Samuel Kotz & Norman L., Johnson: *Limit Distributions of Uncorrelated but Dependent Distributions on the Unit Square*	103
10.	Irini Moustaki: *Latent Variable Models with Covariates*	117
11.	Saralees Nadarajah & Samuel Kotz: *Some New Elliptical Distributions*	129
12.	John Panaretos & Zoi Tsourti: *Extreme Value Index Estimators and Smoothing Alternatives: A Critical Review*	141
13.	Radhakrishna C., Rao, Bhaskara M., Rao & Damodar N., Shanbhag: *On Convex Sets of Multivariate Distributions and Their Extreme Points*	161
14.	Jef Teugels: *The Lifespan of a Renewal*	167
15.	Wolfgang Urfer & Katharina Emrich: *Maximum Likelihood Estimates of Genetic Effects*	179
16.	Evdokia Xekalaki, John Panaretos & Stelios Psarakis: *A Predictive Model Evaluation and Selection Approach—The Correlated Gamma Ratio Distribution*	188
17.	Vladimir M., Zolotarev: *Convergence Rate Estimates in Functional Limit Theorems*	203

Author Index 211
Subject Index 217

List of Contributors

Vic Barnett, Department of Mathematics, University of Nottingham, University Park, Nottingham NG7 2RD, England.
e-mail: vic.barnett@ntu.ac.uk

David Bartholomew, The Old Manse Stoke Ash, Suffolk, IP23 7EN, England.
e-mail: DJBartholomew@compuserve.com

David, R. Cox (Sir), Department of Statistics, Nuffield College, Oxford, OX1 1NF, United Kindom.
e-mail: david.cox@nuffield.oxford.ac.uk

Bradley Efron, Department of Statistics, Stanford University, Sequoia Hall, 390 Serra Mall, Stanford, CA 94305-4065, USA.
e-mail: brad@stat.stanford.edu

Katharina Emrich, Department of Statistics, University of Dortmund, D-44221 Dortmund, Germany

David Freedman, Department of Statistics, University of California, Berkeley, Berkeley, CA 94720-4735, USA.
e-mail: freedman.census@stat.Berkley.EDU

Joe Gani, School of Mathematical Sciences, Australian National University, Canberra ACT 0200, Australia.
e-mail: gani@wintermute.anu.edu.au

Norman, L. Johnson, Department of Statistics, University of North Carolina, Phillips Hall, Chapel Hill, NC, 27599-3260, USA.
e-mail: btrice@stat.unc.edu

Dimitris Karlis, Department of Statistics, Athens University of Economics & Business, 76 Patision St. 104 34, Athens, Greece.
e-mail: karlis@aueb.gr

Leslie Kish, The University of Michigan, USA. [†]

[†] Leslie Kish passed away on October 7, 2000.

Samuel Kotz, Department of Engineering Management and System Analysis, The George Washington University, 619 Kenbrook drive, Silver Spring, Maryland 20902, USA.
e-mail: kotz@seas.gwu.edu

Irini Moustaki, Department of Statistics, Athens University of Economics & Business, 76 Patision St. 104 34, Athens, Greece.
e-mail: moustaki@aueb.gr

Saralees Nadarajah, Department of Mathematics, University of South Florida, Tampa, Florida 33620, USA.
e-mail: snadaraj@chumal.cas.usf.edu

John Panaretos, Department of Statistics, Athens University of Economics & Business, 76 Patision St. 104 34, Athens, Greece.
e-mail: jpan@aueb.gr

Stelios Psarakis, Department of Statistics, Athens University of Economics & Business, 76 Patision St. 104 34, Athens, Greece.
e-mail: psarakis@aueb.gr

Bhaskara M. Rao, Department of Statistics, North Dakota State University, 1301 North University, Fargo, North Dakota 58105, USA.
e-mail: MB_rao@ndsu.nodak.edu

Radhakrishna C. Rao, Department of Statistics, Pennsylvania State University, 326 Thomas Building, University Park, PA, USA 16802-2111, USA.
e-mail: crr1@psu.edu

Damodar, N. Shanbhag, Statistics Division, Department of Mathematical Sciences, University of Sheffield, Sheffield S3 7RH, England.
e-mail: d.shanbhag@sheffield.ac.uk

Jef Teugels, Department of Mathematics, Katholieke Universiteit Leuven, Cerestijnenlaan 200B, B-3030 Leuven, Belgium.
e-mail: Jef.Teugels@wis.kuleuven.ac.be

Zoi Tsourti, Department of Statistics, Athens University of Economics & Business, 76 Patision St. 104 34, Athens, Greece.
e-mail: tsourti@aueb.gr

Wolfgang Urfer, Department of Statistics, University of Dortmund, D-44221 Dortmund, Germany.
e-mail: urfer@omega.statistik.uni-dortmund.de

List of Contributors

Evdokia Xekalaki, Department of Statistics, Athens University of Economics & Business, 76 Patision St. 104 34, Athens, Greece.
e-mail: exek@aueb.gr

Vladimir M. Zolotarev, Steklov Mathematical Institute, Russian Academy of Sciences, Ulitza Vavilova 42, Moscow 333, Russia.
e-mail: zolot@orc.ru

PREFACE

This volume is published in celebration of the 13 years of existence of the Department of Statistics of the Athens University of Economics and Business (www.stat-athens.aueb.gr). The Department was –and still is– the only Department exclusively devoted to Statistics in Greece. The Department was set up in 1989, when the Athens School of Economics and Business was renamed as the Athens University of Economics and Business. Until then, Statistics was part of the Department of Statistics and Informatics. In its 13 years of existence the Department has grown to a center of Statistics in Greece, both applied and theoretical, with many international links. As part of the 13th anniversary celebration, it was decided to put together a volume with contributions from scientists of international calibre as well as from faculty members of the Department.

The goal of this volume is to bring together contributions by some of the leading scientists in probability and statistics of the latter part of the 20th century who are the pioneers in the respective fields. (David Cox writes on "Statistics and Econometrics", C. R. Rao (with M. B. Rao & D. N. Shanbhag) on "Convex Sets of Multivariate Distributions and Their Extreme Points", Bradley Efron on "the Future of Statistics", David Freedman on "Regression Association and Causation", Vic Barnett on "Sample Ordering for Effective Statistical Inference with Particular Reference to Environmental Issues", David Bartholomew on "A Unified Statistical Approach to Some Measurement Problems in the Social Sciences", Joe Gani on "Scanning a Lattice for a Particular Pattern", Leslie Kish on "New Paradigms (Models) for Probability Sampling" (his last paper), Samuel Kotz & Norman L. Johnson on "Limit Distributions of Uncorrelated but Dependent Distributions on the Unit Square", Jef Teugels on "The Lifespan of a Renewal", Wolfgang Urfer (with Katharina Emrich) on "Maximum Likelihood Estimates of Genetic Effects", and Vladimir M. Zolotarev on "Convergence Rate Estimates in Functional Limit Theorems". The volume also contains the contributions of faculty members of the Department. All the papers in this volume appear for the first time in the present form and have been refereed.

Academic and Professional Statisticians, Probabilists and students can benefit from reading this volume because they can find in it not only new developments in the area but also the reflections on the future directions of the discipline by some of the pioneers of the late 20th century. Scientists and students in other scientific areas related to Probability and Statistics, such as Biometry, Economics, Physics and Mathematics could also benefit for the same reason.

The volume is dedicated to professors Constantinos Kevork and Panagiotis Tzorzopoulos who were the first two professors of Statistics of the former Athens School of Economics and Business who joined the newly established Department in 1989. Professor Tzortzopoulos has also served as Rector of the University.

What relates the Department to this volume is that the international contributors, all of them renowned academics, are connected to the Department, one way or another. Some of them (e.g. L. Kish, D. R. Cox, C. R. Rao) have been awarded honorary doctorate degrees by the Department. They, as well as the rest of the contributors, have taught as distinguished visiting professors in the international graduate program of the Department.

I am indebted to all the authors, especially those from abroad, for kindly contributing to this volume but also for the help they have provided to the Department. Finally, I would like to thank Lawrence Erlbaum Publishers for kindly accepting to publish the volume and to make it as widely available as its reputation guarantees.

John Panaretos
Chairman of the Department
Athens, Greece

STOCHASTIC MUSINGS: PERSPECTIVES FROM THE PIONEERS OF THE LATE 20th CENTURY

1
SAMPLE ORDERING FOR EFFECTIVE STATISTICAL INFERENCE, WITH PARTICULAR REFERENCE TO ENVIRONMENTAL ISSUES

Vic Barnett
Department of Computing and Mathematics
Nottingham Trent University, UK

1. Introduction

The random sample is the fundamental basis of statistical inference. The idea of ordering the sample values and taking account both of value and order for any observation has a long tradition. While it might seem strange that this should add to our knowledge, the effects of ordering can be impressive in terms of what aspects of sample behaviour can be usefully employed and in terms of the effectiveness and efficiency of resulting inferences.

Thus, for any random sample $x_1, x_2, \ldots x_n$ of a random variable X, we have the *maximum* $x_{(n)}$ or *minimum* $x_{(1)}$ (the highest sea waves or heaviest frost), the *range* $x_{(n)} - x_{(1)}$ (how widespread are the temperatures that a bridge must withstand) or the *median* (as a robust measure of location) as examples using the ordered sample. The concept of an *outlier* as a representation of extreme, possibly anomalous, sample behaviour or of contamination, also depends on ordering the sample and has played an important role since the earliest days of statistical enquiry. Then again, linear combinations of all ordered sample values have been shown to provide efficient estimators, particularly of location parameters.

An interesting recent development has further enhanced the importance and value of sample ordering. With particularly wide application in environmental studies, it consists of setting up a sampling procedure specifically designed *to choose potential ordered sample values* at the outset– rather than taking a random sample and subsequently ordering it. An example of such an approach is *ranked set sampling* which has been shown to yield high efficiency inferences relative to random sampling. The basic approach is able to be readily and profitably extended beyond the earlier forms of ranked set sampling. We shall review the use of ordered data

- as natural expressions of sample information
- to reflect external influences
- to reflect atypical observations or contamination
- to estimate parameters in models

with some new thoughts on distribution-free outlier behavior, and a new estimator (the *memedian*) for the mean of a symmetric distribution.

2. Inference from the Ordered Sample

We start with the random sample $x_1, x_2 \ldots x_n$ of n observations of a random variable X describing some quantity of, say, environmental interest. If we arrange the sample in increasing order of value as $x_{(1)}, x_{(2)} \ldots x_{(n)}$ then these are observations of the *order statistics* $X_{(1)}, X_{(2)} \ldots X_{(n)}$ from a potential random sample of size n. Whereas the x_i ($i = 1, 2 \ldots n$) are *independent* observations, the order statistics $X_{(i)}, X_{(j)}$, ($i \neq j$) are *correlated*. This often makes them more difficult to handle in terms of distributional behaviour when we seek to draw inferences about X from the ordered sample. (See David, 1981, for a general treatment of ordering and order statistics).

At the descriptive level, the extremes $x_{(1)}$ and $x_{(n)}$, the range $x_{(n)} - x_{(1)}$, the mid-range $(x_{(1)} + x_{(n)})/2$ and the median m (that is, $x_{([n+1]/2)}$ if n is odd, or $(x_{(n/2)} + x_{([n+1]/2)})/2$ if n is even) have obvious appeal and interpretation. In particular the extremes and the median are frequently employed as basic descriptors in exploratory data analysis, and modified order-based constructs such as the *box and whisker plot* utilize the ordered sample as a succinct summary of a set of data (see Tukey, 1977, for discussion of such a non-model-based approach).

More formally, much effort has gone into examining the distributional behavior of the ordered sample values (again David, 1981, gives comprehensive cover). As an example, we have an exact form for the probability density function (pdf) of the range r as

$$g(r) = n(n-1) \int_{-\infty}^{\infty} \{F(x+r) - F(x)\}^{n-2} f(x+r) dF(x)$$

where $f(x)$ is the pdf of X (see Stuart and Ord, 1994, p.494).

But perhaps the most important and intriguing body of work on extremes is to be found in their *limit laws*. Rather like the *Central Limit Theorem* for a sample mean, which ensures convergence to normality from almost any distributional starting point, so we find that *whatever the distribution of X* (essentially), the quantities $x_{(1)}$ and $x_{(n)}$ tend as n increases to approach in distribution one of *only three possible forms*. The starting point for this work is long ago and is attributed by Lieblein (1954) to W. S. Chaplin in about 1860. David (1981, Chapter 9) gives a clear overview of developments and research continues apace to the present time (see, for example, Anderson, 1984; Gomes, 1994).

The three limits laws, are known as, and have distribution functions (df's) in the forms:

1. SAMPLE ORDERING FOR EFFECTIVE STATISTICAL INFERENCE

A: (Gumbel) $F_A(x) = \exp\{-\exp[-(x-\lambda)/\delta]\}$ $-\infty < x < \infty$ $(\delta > 0)$

B: (Frechet) $F_B(x) = \exp\{-[(x-\lambda)/\delta]^{-\alpha}\}$ $x > \lambda$ $(\delta > 0)$

C: (Weibull) $F_C(x) = \exp\{-[-(x-\lambda)/\delta]^{-\alpha}\}$ $x < \lambda$ $(\delta > 0)$

Which of these is approached by $X_{(n)}$ (and $X_{(1)}$ which is simply dual to $X_{(n)}$ on a change of sign) is determined by the notion of *zones of attraction*, although it is also affected by whether X is bounded below or above, or unbounded.

A key area of research is the rate of convergence to the limit laws as n increases – the question of the so-called *penultimate distributions*. How rapidly, and on what possible modelled basis, $X_{(n)}$ approaches a limit law L is of much potential interest. What, in particular, can we say of how the distributions of $X_{(n)}$ stand in relation to each other as n progresses from 40 to 100, 250 or 1000, say? Little, in fact, is known but such knowledge is worth seeking! We shall consider one example of why this is so in Section 3.

Consider the following random sample of 12 daily maximum wind speeds (in knots) from the data of a particular meteorological station in the UK a few years ago:

$$19, 14, 25, 10, 11, 22, 19, 17, 49, 23, 31, 18$$

We have $x_{(1)} = 10$, $x_{(n)} = x_{(12)} = 49$.

Not only is $x_{(n)}$ (obviously) the largest value - the *upper extreme* - but it seems extremely extreme! This is the stimulus behind the study of *outliers*: which are thought of as extreme observations which *by the extent of their extremeness* lead us to question whether they really have arisen from the same distribution as the rest of the data (i.e., from that of X). The alternative prospect, of course, is that the sample is *contaminated* by observations from some other source. An introductory study of the links between *extremes, outliers,* and *contaminants* is given by Barnett (1983) – Barnett and Lewis (1994) provide an encyclopaedic coverage of outlier concepts and methods, demonstrating the great breadth of interest and research the topic now engenders.

Contamination can, of course, take many forms. It may be just a reading or recording error – in which case *rejection* might be the only possibility (supported by a *test of discordancy*). Alternatively, it might reflect low-incidence mixing of X with another random variable Y whose source and manifestation are uninteresting. If so, a robust inference approach which draws inferences about the distribution of X while *accommodating* Y in an uninfluential way might be what is needed. Then again, the contaminants may reflect an exciting unanticipated prospect and we would be anxious to *identify* its origin and probabilistic characteristics if at all possible. *Accommodation, identification,* and *rejection* are three of the approaches to outlier study, which must be set in terms of (and made conditional on) some model F for the distribution of X. This is so whether we are examining univariate data, time series, generalized linear model outcomes, multivariate observations, or whatever the base of our outlier interest within the rich field of methods now available.

But what of our extreme daily wind speed of 49 in the above data? We might expect the wind speeds to be reasonably modelled by an extreme value distribution – perhaps of type B (Frechet) or A (Gumbel), since they are themselves maxima over a 24-hour period. Barnett and Lewis (1994, Section 6.4.4) describe various statistics for examining an upper outlier in a sample from a Gumbel distribution. One particular test statistic takes the form of a Dixon statistic,

$$(x_{(n)} - x_{(n-1)})/(x_{(n)} - x_{(1)}).$$

For our wind-speed data with $n=12$, this takes the value $\frac{18}{39} = 0.46$ which according to Table XXV on page 507 of Barnett and Lewis (1994) is not significant. (The 5% point is 0.53, so notice how critical is the value of $t_{(n-1)}$ i.e., $t_{(11)}$. If instead of 31 it were 28, then $x = 49$ would have been a *discordant outlier* at the 5% level. This illustrates dramatically how some outlier tests are prone to 'masking': Barnett & Lewis, 1994, pp. 122-124.) Thus we conclude that although 49 seems highly extreme it is not extreme enough to suggest contamination (e.g., as a mis-reading or a mis-recording or due to freak circumstances).

A fourth use of ordered data is in regard to basic estimation of the parameters of the distribution F followed by X. Suppose X has df which takes the form $F[(x - \mu/\sigma]$ where μ reflects location and σ scale or variation. If X is symmetric, μ and σ are its mean and standard deviation. Nearly 50 years ago, Lloyd (1952) showed how to construct the BLUE or *best linear unbiased estimator* of μ and of σ based on the order statistics, by use of the Gauss-Markov theorem.

Suppose we write $U_{(i)} = (X_{(i)} - \mu)/\sigma$ $(i = 1, 2, \ldots, n)$ as the *reduced (standardised) order statistics* and let α and V denote the mean vector and variance covariance matrix of U. Note that V is *not* diagonal since the $U_{(i)}$ and $U_{(j)}$ (for $i \neq j$) are correlated. Using the principle of extended least squares we obtain the BLUE θ^* of θ - where $\theta' = (\mu, \sigma)$ - in the form

$$\theta^* = (\mathbf{A}' \mathbf{V}^{-1} \mathbf{A})^{-1} \mathbf{A}' \mathbf{V}^{-1} \mathbf{x}$$

with variance/covariance matrix

$$\text{var}(\theta^*) = \sigma^2 (\mathbf{A}' \mathbf{V}^{-1} \mathbf{A})^{-1}$$

where $\mathbf{A} = (\mathbf{1}, \alpha)$ with $\alpha' = \{\alpha_i\} = \{E(U_{(i)})\} = \{E[(X_{(i)} - \mu)/\sigma]\}$ and $\mathbf{V} = \{\upsilon_{ij}\}$ is the variance/covariance matrix of the reduced order statistics $U_{(i)} = (X_{(i)} - \mu)/\sigma$.

This can be readily separated to yield the individual BLUE's, μ^* and σ^*. (See David, 1981, for broader discussion of optimal and sub-optimal estimators based on order statistics and of how they compare with estimators based on the unordered sample.)

This approach is central to the more modern environmentally important principles of *ranked set sampling*, which we consider briefly in Section 3.

3. Possible New Routes for Outliers and for Order-Based Samples

Some of the principles reviewed suggest possible developments in outlier methodology on the one hand and in order-based estimation on the other.

3.1 A Distribution-Free Approach to Outliers

It is clear from the above outline, that the methodology of outliers depends crucially on the form of the null (no-contamination) model. Thus, for example, even a discordancy test of a single upper outlier $x_{(n)}$ based on the statistic $t = (x_{(n)} - x_{(n-1)}) / (x_{(n)} - x_{(1)})$ is constrained in this way – since the null distribution of t (and its critical values) depends *vitally* on the form of F. The distribution of t and its percentage points will obviously be different if F is normal, exponential, Gumbel, etc. Yet we may not have any sound basis for assuming a particular form of F, especially if the only evidence is the single random sample in which we have observed an outlier. *This is the fundamental problem of outlier study.*

In practice, this dilemma is well-recognized and is usually resolved by a judicious mix of historical precedent, broad principle, association and wishful thinking (as in all areas of model-based statistics).

Thus it may be that a previous related study, and general scientific or structural features of the practical problem, link with formal statistical considerations (e.g., the Central Limit Law, characteristics of extremal processes) to support a form for F, such as a normal distribution or an exponential distribution. We then relate our inferences to appropriate null (no-outlier) distributions for that particular F.

But we are concerned, of course, in studying outliers which as *extremes* must have the *distributional behaviour of extremes*, which we have just seen to be essentially distributionally independent of the family F from which the sample has been chosen – in view of the limit laws. So in principle it seems that we might essentially *ignore F and examine outlier behavior in terms of properties of the extreme value distribution which is being approached by $x_{(1)}$ or $x_{(n)}$* (or by some appropriate outlier function of them). This is an attractive prospect: a *distribution-free* and hence *highly robust alternative to the usual model-based methods*.

So what is the difficulty? Precisely the following. Although $x_{(n)}$ approaches *A, B* or *C*, we have to deal with *finite samples* and not enough is known in detail about *how quickly and in what manner* the forms *A* or *B* or *C* are approached as *n* progresses from, say 40 to 100 to 250, etc. The study of convergence to the limit laws and of 'penultimate distributions' is not yet sufficiently refined for our purpose (See Gomes, 1994, for some of the latest developments).

To consolidate this point consider Table 1.1 which shows samples of maximum daily, 3-daily, weekly, fortnightly, and monthly wind speeds (in knots) at a specific location in the UK.

Daily
36 46 13 18 34 19 23 15 18 14 28 10 31 28 22 40 20 23 28
3-Daily
33 31 21 19 22 28 29 25 36 41 16 24 43 20 38 51 34 20 31
Weekly
40 36 47 21 41 27 34 32 45 42 54 19 30 31 24 31 33 34 36
Fortnightly
35 32 45 37 39 31 34 28 47 58 31 33 51 42 50 47 40 41 52
Monthly
40 44 39 32 48 36 51 40 38 52 62 51 39 50 42 56 29 36 45

Table 1.1: Samples of maximum windspeeds (in knots) at a single UK location over days, 3-days, weeks, fortnights, and months.

Assuming (reasonably) that these approach the limit law A (they are all maxima) we would expect to find that plots of $\ln \ln[(n + 1)/i]$ against $x_{(i)}$ in each case, will yield approximately linear relationships. It is interesting to confirm from the data that this is indeed so. We will further see that we obtain the natural temporal ordering we would expect (reflected, particularly, in the implied differences in the values of, particularly, λ) in the approximating extreme value distribution in each case. Essentially the plots are parallel with intercepts *increasing with the lengths of the periods over which the maximum is taken*.

Davies (1998) also carried out an empirical study of limiting distributions of wind speeds (again from a single UK site) and fitted Weibull distributions to maxima over days, weeks, fortnights, months, and 2-month periods. Figure 1 (from Davies, 1998) shows the fitted distributions in which the time periods over which the maxima are taken increase monotonically as we move from the left-hand distribution to the right-hand one.

We need to know much more about how the distributions change with change in the maximizing period. It might be hoped that we can obtain a clearer understanding of how the limit distribution of an extreme is approached in any specific case as a function of sample size n and that such knowledge might eventually lead to an essentially new (largely) distribution-free outlier methodology.

1. SAMPLE ORDERING FOR EFFECTIVE STATISTICAL INFERENCE   7

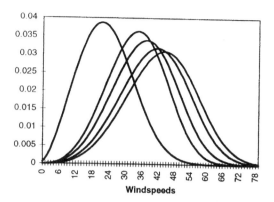

Figure 1.1: Fitted Weibull distributions to maximum windspeeds over different periods of time from 1 day to 2 months.

3.2 The Median and the Memedian

Ranked set sampling has become a valuable method particularly in environmental study. Barnett (2000) remarked:

"A method which is becoming widely used for sampling in the context of measuring expensive environmental risk factors is that of *ranked set sampling*. It can be used for the estimation of a mean, a measure of dispersion, quantiles or even in fitting regression models. The gains can be dramatic: efficiencies relative to simple random sampling may reach 300%.

The aim is to employ concomitant (and cheaply and readily available, sometimes subjective) information to seek to 'spread out' the sample values over their possible range. This can result in a dramatic increase in efficiency over simple random sampling. The method has been around for nearly 50 years since it was first mooted in an agricultural/environmental context (McIntyre, 1952). Further modifications continue apace to improve efficiency and applicability for different distributional forms of the underlying random variable and of the type of inference needed."

The method works as follows (Barnett, 2000):
"Conceptual random samples, of observations of the random variable X, take the form

$$\begin{array}{cccccc} x_{11} & \ldots & x_{21} & \ldots & x_{1n} \\ x_{21} & \ldots & x_{22} & \ldots & x_{2n} \\ \\ x_{n1} & \ldots & x_{n2} & \ldots & x_{nn} \end{array}$$

From each subsample we take one measured observation $x_{i(i)}$: the ith ordered value in the ith sample ($i = 1, 2, \ldots, n$). The ranked-set sample is then defined as $x_{1(1)}, x_{2(2)}, \ldots, x_{n(n)}$. In early applications, the mean μ of the underlying distribution was estimated by

$$\bar{\bar{x}} = \Sigma x_{i(i)}/n \qquad (1)$$

which is the *ranked set sample mean*, which compares favorably with $\bar{x}$ (the mean of a random sample of size n; not of size n^2, because measurement is assumed to be of overriding effort compared with ordering). We find that $\bar{\bar{x}}$ is unbiased and that (for $n > 2$) typically

$$\text{var}(\bar{\bar{x}}) < \text{var}(\bar{x})$$

often markedly so, for different sample sizes and distributions, if we have correctly ordered the potential observations in each conceptual subsample."

It will prove interesting to extend (1) to a more general form: that of an arbitrary linear combinations of the $x_{i(i)}$ terms. We consider estimators of the form

$$\mu^* = \Sigma \gamma_i x_{i(i)} \qquad (2)$$

In the general case where X has df $F[(x - \mu)/\sigma]$, we just noted how to determine the BLUE of μ and σ from the ordered sample. For the ranked set sample $x_{1(1)}, x_{2(2)} \ldots, x_{n(n)}$ we have a simplification in that the variance covariance matrix is now *diagonal* (since $X_{i(i)}, X_{j(j)}$, are independent if $i \neq j$) and V in the development of Section 2 can be replaced with $W = \text{diag}(v_{ii}) = \text{diag}(v_i)$. So if we write the optimal estimators as

$$\mu^* = \sum_{j=1}^{n} \gamma_i x_{i(i)}, \qquad \sigma^* = \sum_{j=1}^{n} \eta_i x_{i(i)}$$

we have

$$\gamma_i = \frac{(1/v_i)\left[\sum_{j=1}^{n}(\alpha_j^2/v_j) - \alpha_i \sum_{j=1}^{n}(\alpha_j/v_j)\right]}{\Delta} \qquad (3)$$

$$\eta_i = \frac{(1/v_i)\left[\alpha_i \sum_{j=1}^{n}(1/v_j) - \sum_{j=1}^{n}(\alpha_j/v_j)\right]}{\Delta} \qquad (4)$$

1. SAMPLE ORDERING FOR EFFECTIVE STATISTICAL INFERENCE

where

$$\Delta = \sum\left(\frac{\alpha_j^2}{\upsilon_j}\right)\sum\left(\frac{1}{\upsilon_j}\right) - \left[\sum\left(\frac{\alpha_j}{\upsilon_j}\right)\right]^2$$

with

$$\left.\begin{array}{l}\mathrm{var}(\mu^*) = \sigma^2 \dfrac{\sum_{i=1}^{n}(\alpha_i^2/\upsilon_i)}{\Delta} \\[2ex] \mathrm{var}(\sigma^*) = \sigma^2 \dfrac{\sum_{i=1}^{n}(1/\upsilon_i)}{\Delta}\end{array}\right\} \quad (5)$$

The properties of these, and related estimators, are discussed by Barnett and Moore (1997), Sinha et al. (1996), Stokes (1995) and Barnett (2000). In particular μ^* is highly efficient in comparison with the random sample mean (from an unordered sample) and more efficient than the ranked set sample mean.

Modified schemes in which we take different numbers of observations of different $x_{i(i)}$ have been discussed in terms of sampling design and effect by Kaur et al. (1997) and Barnett (1999).

Suppose we consider an extreme version of such a differential choice of the $x_{i(i)}$: namely that from each conceptual sample we chose only the median m_i. So our sample is now the set of n values m_i ($i = 1, 2, \ldots n$). Rather than spreading out the sample – the original aim of ranked set sampling – we have now concentrated it into all the medians. Could this be sensible for estimating μ in a symmetric distribution where X has df $F[(x - \mu/\alpha)]$?

Suppose the median m has variance $v_{(m)}\sigma^2$. Then, if we define the *memedian* M to be the mean value of the medians:

$$M = \frac{1}{n}\sum_{1}^{n} m_i$$

its sampling variance will be $v_M = v_{(m)}\sigma^2/n$ where $v_{(m)}\sigma^2$ is the variance of an individual sample median, obtained from the diagonal variance covariance matrix W. In comparison, we know that the ranked set sample mean, $\bar{\bar{x}}$, has variance $v_{\bar{x}} = \sum_{1}^{n} v_{ii}\sigma^2/n^2$ so that the relative efficiency of M and $\bar{\bar{x}}$ is $e_1 = \sum v_{ii}/(nv_{(m)})$. Clearly M will be more efficient than $\bar{\bar{x}}$ if $\bar{v}_{ii} \geq v_{(m)}$ which will be true if $v_{(m)} = min\ \{v_{ii}\}$. Can this happen? *We will see that it can.*

For illustrative purposes, we show in Table 1.2 the variances of standardized order statistics from samples of size 5 for four symmetric distributions which in standardized forms have pdf's as follows:

- Normal $\exp(-x^2/2)$
- Uniform   1
- Triangular $4x+2$ $(-1/2 < x < 0)$
                   $2-4x$ $(0 < x < 1/2)$
- Double exponential $\exp\{-|x|\}$

We see, not unsurprisingly, that the variances of the standardized order statistics are symmetric about that of the median, but *what is perhaps surprising is that sometime the median has largest variance, sometimes the smallest.* (Results for the triangular and double exponential distributions come from Sarhan, 1954).

Distribution	v_{ii}				
Normal	.4475	.3115	.2868	.3115	.4475
Uniform	.01984	.03175	.03571	.03175	.01984
Triangular	.1524	.1407	.1333	.1407	.1524
Double exponential	1.4703	.5025	.3512	.5025	1.4703

Table 1.2: Variances of standardized order statistics for samples of size 5 from symmetric distributions

So for the normal, triangular and double exponential distributions the *memedian* is more efficient than the ranked set sample mean. That this is not universally true is seen from the results for the uniform distribution. The values of the relative efficiency e_1 in the four cases are:

$$1.67, 0.78, 1.080, 2.45$$

showing major efficiency gains for the normal and double exponential distributions.

Is it even possible that M is more efficient than the ranked set BLUE, μ^*? The relative efficiency of M and μ^* is now $e_2 = n/[v_{(m)} \sum_1^n (1/v_{ii})]$.

Again, we can consider this for the four distributions in Table 1.2 illustrated for sample size $n = 5$. The values of e_2 are now

$$1.21, 0.73, 1.077, 1.74$$

so that again (for the same three cases) we conclude *rather surprisingly* that M can indeed be more efficient than μ^*. Further, it is much easier to calculate and we recall that it is also (in appropriate distributional circumstances) *a fortiori* more efficient than the ranked set sample mean, $\bar{\bar{x}}$. Clearly, we would need to pursue these results in more detail, for different sample sizes and for different distributions.

Accordingly, it would be useful to extend the calculations and general distributional assumptions to seek to clarify just when, and to what extent, the *memedian* is to be preferred to the ranked set sample mean or to the 'optimal' ranked set sample BLUE.

4. Discussion

We have reviewed the different ways in which the ordered sample is important in statistical inference and in the process have identified two new approaches of potential value
- a possible distribution-free method for handling outliers
- the *memedian* as a possible improved estimator of the mean of a symmetric distribution

both of which clearly merit further study.

5. References

Anderson, C. W. (1984). Large deviations in extremes, In J. Tiago de Oliveira (1984) (ed.), *Statistical Extremes and Applications*, D. Reidel, 325-340.
Barnett, V. (1983). Principles and methods for handling outliers in data sets. In T. Wright, (ed.) *Statistical Methods and the Improvement of Data Quality*. Academic Press, Orlando.
Barnett, V. (1999). Ranked set sample design for environmental investigations, *Environmental and Ecological Statistics*, 6, 59-74.
Barnett, V. (2000). Statistically evaluating risk in environmental issues. *Proceedings of the 1999 Portuguese Statistical Society Congress*, Porto, Portugal.
Barnett, V. and Lewis, T. (1994). *Outliers and Statistical Data* 3rd ed., Wiley Chichester.
Barnett, V. and Moore, K.L. (1997) Best linear unbiased estimators in ranked-set sampling with particular reference to imperfect ordering, *Journal of Applied Statistics*, 24, 697-710.
David, H. A. (1981). *Order Statistics*, 2nd ed., John Wiley, New York.
Davies, G. (1998). Extreme Value Theory, *Undergraduate Project*, School of Mathematics, University of Nottingham.
Gomes, M. I. (1994). Penultimate behaviour of the extremes. In Galambos, J. *et al* (1994) (eds) *Extreme Value Theory and Applications*, Kluwer, Amsterdam.
Kaur, A. Patil, G. P. and Taillie, C. (1997). Unequal allocation models for ranked set sampling with skew distributions, *Biometrics*, 53, 123-130.
Lieblein, J. (1954). Two early papers on the relation between extreme values and tensile strength, *Biometrika*, 41, 559-560.
Lloyd, E. H. (1952). Least-squares estimation of location and scale parameters using order statistics, *Biometrika*, 39, 88-95.
McIntyre, G. A. (1952). A method of unbiased selective sampling using ranked sets. *Australian Journal of Agricultural Research*, 3, 385-390.
Sarhan, A. E. (1954). Estimation of the mean and standard deviation by order statistics, *Ann Math Statist*, 25, 317-328.
Sinha, B. K., Sinha, R. K. and Purkayastha, S. (1996). On some aspects of ranked set sampling for estimation of normal and exponential parameters. *Statistical Decisions*, 14, 223-240.
Stokes, S. L. (1995). Parametric ranked set sampling. *Annals of the Institute of Statistical Mathematics*, 47, 465-482.

Stuart, A. and Ord, K. (1994). *Kendall's Advanced Theory of Statistics Vol 1 Distribution Theory* 6th ed., Edward Arnold, London.

Tukey, J. W. (1977). *Exploratory Data Analysis*, Vol 1, Addison-Wesley, Reading, Mass.

Received: December 1999

2
A UNIFIED STATISTICAL APPROACH TO SOME MEASUREMENT PROBLEMS IN THE SOCIAL SCIENCES

David Bartholomew
London School of Economics, UK

1. Introduction

Measurement is the cornerstone of science and if social science is to justify its name it must aim for standards of measurement which bear comparison with those of natural science. In some cases there is little problem in meeting this requirement. For example, birth rates, life expectancy, and average hospital waiting times are well-defined concepts, which can be directly measured from readily available data. However, there is another class of concepts which are regarded as quantitative and yet which cannot be directly and unambiguously measured. Quality of the environment is such an example. We use the term in social debate in a way that implies it is something which one can have more or less of. If pressed to justify statements of this kind we would point to a whole collection of directly observable quantities such as levels of atmospheric pollutants, water quality, noise levels, contaminants in food, and so on. It thus appears that the term "quality of the environment" is a shorthand for something to which a constellation of other observable variables are assumed to contribute. It is, in short, a collective property of a set of variables. The problem is how to extract, in some sense, and then to combine into a single measure what each variable is contributing. The statistical problem is to provide a theoretical framework within which this can be done.

This is not a new problem and work on it goes back to the beginning of the 20th century. In the educational and psychological fields the modern subject of psychometrics represents the most highly developed strand of this tradition. Recently there has been a growing interest in measurement in sociology and, especially, in the relationships which exist between such measures. What has been lacking until recently is a unified approach to this class of measurement problems. This has meant that development in various fields has failed to benefit from closely related work, with the result that there has been duplication of effort and, in one case at least, inconsistent conclusions have been reached.

The purpose of this chapter is to draw attention to recent work on a unifying framework and then to show how its use illuminates a number of obscure and controversial topics in the measurement of social variables.

2. The Latent Variable Framework

To focus our thinking we take a very simple and familiar example from educational testing. Children vary in their ability to perform tasks such as basic arithmetic. This gives rise to the notion of "arithmetical ability" as something which individuals have in varying degrees. In order to measure this ability, tests are administered in which, for example, a child might be asked to do 20 simple sums which are scored as "right" or "wrong." The total number correct seems to be a sensible summary measure of ability and it is commonly used in this way. On reflection we might wonder whether some other summary measure, such as the geometric mean, might not be better. Or, recognizing that the test items might not be of equal difficulty, whether some items should be given more weight than others. It is questions of this kind to which we might expect a theoretical framework to provide answers.

The essence of the approach is to add a random variable, representing the unobservable quantity, to the set of those corresponding to the observed variables (also known as indicators). The additional random variable is known as a latent variable. Once this is done the question of how to measure the latent quantity can be posed in terms of its conditional distribution. This approach, in rudimentary form, goes back at least to Dolby (1976) but is given in its most complete form to date in Bartholomew (1996) and Bartholomew and Knott (1999). Here we shall briefly summarize the basic essentials sufficient for what follows.

Let there be p observable random variables denoted by $x' = (x_1, x_2, ..., x_p)$ and one latent variable denoted by z. (Much of what follows holds if z is vector-valued and, later, we shall need this extension). An important feature of our general formulation is that the x's and z may be categorical, discrete, or continuous or, in the case of x, a mixture. We shall henceforth use the term *metrical* to include continuous and discrete variables.

Since z is unobservable, the only distribution we can make inferences about is $f(x)$, the joint distribution of x. This may be expressed as

$$f(\mathbf{x}) = \int f(\mathbf{x} \mid z) dF(z) \qquad (1)$$

where the integral is over the range of z. It is immediately clear that one cannot uniquely determine either $f(x|z)$ or the prior distribution of z. Hence we cannot determine the posterior distribution of z which is given by

$$dF(z \mid \mathbf{x}) = f(\mathbf{x} \mid z) dF(z) / f(\mathbf{x}) \qquad (2)$$

We thus appear to be at an impasse but further progress could be made if $f(x|z)$ happened to have the form

$$f(\mathbf{x} \mid z) = a(X, z) b(\mathbf{x}) c(z) \qquad (3)$$

where X is a scalar function of x. In that case we would be able to write

2. A UNIFIED STATISTICAL APPROACH

$$dF(z \mid \mathbf{x}) = \frac{a(X,z)b(\mathbf{x})c(z)dF(z)}{b(\mathbf{x})\int a(X,z)c(z)dF(z)} = \frac{a(X,z)c(z)dF(z)}{e(X)} \qquad (4)$$

This would imply that the conditional distribution of z given $\mathbf{x}$ depended on $\mathbf{x}$ only through the scalar quantity X. In that sense, therefore X would then contain all the information in $\mathbf{x}$ relevant to z. It follows that any "measure" of z should be a function of X. If z is metrical one could predict z given X by $E(z|X)$ or some other measure of location. If z were a binary (0/1) variable coding two latent classes then $E(z|X)$ would be the posterior probability of being in the class coded 1. In a certain sense X may be described as "sufficient" for z; like a sufficient statistic it contains all the information in $\mathbf{x}$ about z.

For practical purposes the usefulness of the foregoing analysis depends on whether one can justify the choice of a model for which the factorization of (3) is possible. It turns out that this can be done in a wide enough range of circumstances to cover most practical needs. In particular, if the x's are conditionally independent, that is if

$$f(\mathbf{x} \mid z) = \prod_{i=1}^{p} f_i(x_i \mid z) \qquad (5)$$

and if $f_i(x_i|z)$ is a member of the one-parameter exponential family ($i=1,2,...,p$).

The conditional independence of (5) is necessary because it implies that no latent variable, other then z, is required to account for the dependencies among the x's. If (5) did not hold, we could infer that at least one other latent variable was needed. The membership of the exponential family, in addition to (5), then ensures the factorization given in (3). Since the Bernoulli, multinomial, and normal distributions are all included within this family, categorical and normally distributed variables are included. We emphasize again that the x's do not need to have the same distribution.

If z is categorical we have a latent class problem and the posterior distribution will tell us the probability that an individual with a given $\mathbf{x}$ will fall into the latent class indicated by z. This represents what is often called a nominal scale of measurement. If z is continuous there is no empirical basis for anything higher than an ordinal level of measurement. This is because the calculation of any measure of location of the posterior distribution depends on the prior $f(z) = dF(z)/dz$ which is, as we noted, indeterminate. We can only estimate it if we make some assumption about the form of $f_i(x_i|z)$ and, even then, it may be difficult to estimate $f(z)$ with any precision. Different choices of $f(z)$ will lead to different choices of the function of X to be used. It may be shown that $E(z|X)$ is a monotonic function of X, whatever $f(z)$, so that all $f(z)$'s will lead to the same estimated rankings of the individuals. This is, in fact, what we should expect because if z cannot be directly observed there is no natural scale to calibrate it.

The formulation we have given includes virtually all existing models. As we have noted, if z is an indicator variable we have latent class analysis. When z

is continuous and when the conditional distributions of x_i given z are normal we have the factor model. If the x's are binary the latent trait model emerges. Here we shall not look at these important special cases because our purpose is to emphasize results which apply to all members of the class.

3. Measurement Issues in a General Perspective

Here we take a number of measurement issues on which new light is thrown by the approach just outlined.

3.1 Factor Scores

There is a longstanding and highly controversial literature in the factor analysis field on what are called *factor scores*. A recent and illuminating example of the debate was initiated by Maraun (1996). The problem is that of how to measure z; that is to find some function of the x's to 'estimate' or 'predict' z. Within the framework we have adopted here the matter is entirely straightforward. After x is observed our knowledge about z is contained in the conditional distribution of z given x. The value of z is indeterminate in the sense that there is no single value that we can assign to it. Rather our uncertainty is expressed by a probability distribution. The best we can do is to use some measure of location of the distribution as a score for z. The expectation, $E(z|X)$, is an obvious choice and this happens to coincide with one of the traditional factor scores derived by other methods. But, as we have seen, it depends on the arbitrary choice of a prior distribution. In many cases it is intuitively appealing to use X itself since it often turns out to be a linear combination of the x's.

It is interesting that there has been no such controversy in the case where z is categorical. This illustrates how little cross-fertilization there has been between the various branches of latent variable modeling. "Measurement" here consists of identifying a latent class to which an individual belongs. This has always been done by finding the posterior probability distribution over the latent classes, which is precisely what our general strategy indicates. The general approach is not, however, restricted to these special cases. It works whatever kinds of variables are included among the x's and, incidentally, whether or not their conditional distributions belong to the exponential family.

3.2 Reliability

Since we cannot determine z precisely we need some way of indicating the uncertainty associated with our prediction. This is what a measure of reliability is designed to do. If our knowledge of z is conveyed by its posterior distribution, then our uncertainty about it can be measured by the posterior variance, $var(z|X)$, or standard deviation. As we conventionally take the prior to have unit variance, it follows that $var(z|X) \leq 1$. Hence we may take

$$r^* = 1 - var(z|X)$$

2. A UNIFIED STATISTICAL APPROACH

as a measure of reliability since reliability is complementary to dispersion. The smaller $var(z|X)$, the larger the reliability and conversely. Like the mean, r^* depends upon the choice of prior but it may be shown that the dependence is slight if p, the number of items, is reasonably large. In practice it appears that the posterior distribution is usually close to normal, so the mean and variance together provide a complete description of the distribution. Further details will be found in Bartholomew (1996).

This measure is not the same as that which has been traditionally used in those parts of latent variable analysis which have used the concept. The usual procedure is to base the measure on the distribution of X given z. Thus if S is the chosen function of X we would compare $var(S|z)$ with $var(S)$, the argument being that if S is a good predictor of z then fixing z will reduce the variance of S. Consequently we could measure reliability by $\{1-var(S|z)/var(S)\}$. Like r^* this would be 1 if S determines z exactly (when $var(S|z) = 0$) and is 0 if S conveys no information about z. In the form we have defined it z is unknown so the reliability could not be determined. Instead, therefore, we have to take the average value given by

$$r = 1 - \frac{Evar(S \mid z)}{var(S)} \qquad (6)$$

Our proposed measure, r^*, is a function of X but this is known. In practice it turns out that the dependence on X is slight except possibly for extreme values of X.

Rather remarkably the two measures r and r^* can be shown to be close and, in one important case, are identical. Nevertheless r^* has a more convincing rationale in that it is calculated on the basis of what is known, namely X. For many purposes the posterior standard deviation $\sigma(z|X)$ is more directly interpretable and is, of course, equivalent to r^*. Here again the general approach serves to unify and generalize an important measurement concept.

3.3 The Adequate Treatment of Categorical Data

In much social research, especially arising from sample surveys, the observed variables are a mixture of continuous and categorical. Hitherto there have been two ways of handling this situation. One is to downgrade the level of measurement by reducing all metrical variables to categorical, often binary, variables. This provides a valid but less efficient method because information is lost. The other method is to upgrade the categorical variables to metrical variables. This can only be done by adding an assumption which is, usually, that the categorical variables have resulted from categorizing a continuous variable by dividing its range into segments and merely recording into which segment a sample member falls. In other words a categorical variable is regarded as an incompletely observed continuous variable.

If we make the further assumption that all pairs of these continuous variables have normal bivariate distributions the product moment coefficients can be estimated as tetrachoric, polychoric, or bi-serial correlations according to the level of measurement of the variables involved. This enables us to carry out a factor analysis although there may be technical problems if, for example, the correlation matrix happens not to be positive definite. Any attempt to compute standard errors will be invalid. Other problems with this approach are discussed in Lee and Poon (1999) in relation to the software package LISCOMP. The new approach allows us to treat the data as they are, without loss of information on the one hand, or the introduction of arbitrary information, about the underlying distribution, on the other. Further details will be found in Bartholomew and Knott (1999, chapter 7) and Moustaki (1996).

4. Comparisons

In all of the discussion so far we have been concerned with constructing a scale of measurement for a single latent variable. Having constructed such a scale we may wish to compare different populations. In particular, to compare the mean values of z in two populations, say. This often arises in practice when comparing average levels of attainment in different schools or age groups. This poses formidable problems. We recall from Section 2 that any one-to-one transformation of z in (1) leaves $f(\mathbf{x})$ unchanged. In particular any linear transformation of z will have this result. Hence the mean and the variance of the prior distribution is arbitrary and we are free to fix them as we will. As a matter of convention we used a standard normal distribution with mean zero and variance one. Our z-scores were therefore determined relative to that prior distribution. In computing z-scores for two populations, both sets are based on a population mean of zero, and therefore there is no means of comparing the populations since we have removed any existing difference before calculating the scores. If there actually is a difference between the population means, then that difference will be absorbed into the parameters of the conditional distributions $\{f_i(x_i|z)\}$. If we are prepared to assume that the value of x_i observed for a given value of the unstandardized latent variable is the same in both populations, then any difference in the location of the populations will be reflected in the estimates of these parameters. Differences observed in them can then be interpreted as the effects of population differences.

Although this result is perfectly general it is most easily seen in the case of the linear factor model. We write this as

$$x_i = \mu_i + \alpha_i Z + e_i \quad (i=1,2,\ldots,p)$$

where $Z \sim N(n, \sigma^2)$ and $e_i \sim N(0, \psi_i)$. The e's and Z are mutually independent. Suppose, now, we have two populations distinguished by the additional subscripts 1 and 2. Then let

$$z_j = (Z - n_j)/\sigma_j \quad (j=1,2).$$

For the two populations we shall then have

$$\left.\begin{array}{l}x_{i1} = \mu_i + \alpha_i(n_1 + \sigma_1 z_1) + e_{i1} \\ x_{i2} = \mu_i + \alpha_i(n_2 + \sigma_2 z_2) + e_{i2}\end{array}\right\} (i = 1,2,\ldots, p) \qquad (7)$$

Our assumption that Z has the same effect on the x's in both populations implies that μ_i, α_i and ψ_i are the same in both and so do not need suffixes. But when we fit the model to each population with a standardized z the new 'α's' will be $\alpha_i\sigma_2$ and $\alpha_i\sigma_1$ which will not be the same unless $\sigma_1 = \sigma_2$. Any difference is therefore indicative of a difference in population standard deviations. To get at the difference of population means we need to know that X for this model is

$$X_j = \sum_{i=1}^{p} \frac{\alpha_i}{\psi_i}(x_{ij} - \mu_i) \qquad (j=1,2) \qquad (8)$$

It is clear from (7) that
$$E(x_{i1} - x_{i2}) = \alpha_i(n_1 - n_2).$$

Hence if X_1 is significantly greater than X_2 we may infer that $n_1 > n_2$.

It appears, therefore, that comparisons can be made of the means and dispersions of two populations but only if we are prepared to make an empirically unverifiable assumption. Without that assumption any apparent population differences will be confounded with differences in the way the x's are related to the latent variable in the two populations.

5. References

Bartholomew, D.J. (1996). *The Statistical Approach to Social Measurement*, Academic Press, San Diego, CA.

Bartholomew, D.J. & Knott, M. (1999). *Latent Variable Models and Factor Analysis*, 2nd edition, Arnold, London.

Dolby, G.R. (1976). Structural Relations and Factor Analysis. *Mathematical Scientist*, (Supplement No. 1), 25-29.

Lee, S.-Y. & Poon, W.-Y. (1999). Two Practical Issues in Using LISCOMP For Analyzing Continuous and Ordered Categorical Variables. *British Journal of Mathematical and Statistical Psychology*, 52, 195-211.

Maraun, M.D. (1996). Metaphor Taken as Math: Indeterminacy in the Factor Analysis Model. *Multivariate Behavioral Research*, 31, 517-538.

Moustaki, I. (1996). A Latent Trait and a Latent Class Model for Mixed Observed Variables. *British Journal of Mathematical and Statistical Psychology*, 49, 313-334.

Received: December 1999

3
SOME REMARKS ON STATISTICAL ASPECTS OF ECONOMETRICS

D.R.Cox
Department of Statistics
Nuffield College, Oxford, UK

1. Introduction

This chapter gives some miscellaneous comments from the outside viewpoint of a statistician on the challenges associated with quantitative methods in econometrics. There are several matters to be considered and the relative importance of these must depend on the particular context. Emphasis can be placed on

1. the provision of better data and of data more specific to the research questions of concern, including detailed discussion of measurement issues and data definition and especially data quality;
2. more intensive analysis by already developed methods of currently available data;
3. more critical interpretation of analyses made by currently available methods;
4. the development and deployment of more elaborate methods of analysis;
5. the deeper incorporation into analysis of specific subject-matter concepts and theory.

All these are surely important and some are complementary rather than contradictory. In particular the last two issues are quite strongly connected; more incorporation of specific quantitative considerations is likely to lead to new types of model in turn needing special methods for their analysis. While the development of new methods is the aspect of most immediate interest to the theoretical statistician, there is some danger of overweighting this aspect! In particular, issues of data quality and relevance, while underemphasized in the theoretical statistical and econometric literature, are certainly of great concern in much statistical work.

Of central importance in much statistical thinking is the explicit recognition that conclusions are uncertain and that this uncertainty, or at least that part of it that arises from haphazard variation in the data, should be measured. There is moreover often an implicit tradition, stemming perhaps from R.A. Fisher's work largely in biometry, that each investigation should be self-contained, providing its own estimate of error. At the same time it is clear that interpretation cannot proceed in isolation and the merging of the results of different investigations, and especially investigations of different types, is crucial. This underlies the fifth point and some of the criteria to be discussed in the next section.

While every individual field of application has its own specific difficulties many of the issues are of very broad relevance in the observational and experimental sciences and their associated technologies.

This chapter in particular sketches some of the areas of statistical research that are especially active at the moment; in most but not all there is parallel work in the econometric literature.

2. Criteria for Statistical Models

The analysis of data via tables and graphs is very important especially, but not only, for the presentation of conclusions. Nevertheless we concentrate here on methods with an explicit base in a probabilistic model. Much formal statistical discussion takes a family of models as given and develops the general concepts required for its analysis. Yet it seems clear that the initial choice of a family of models as a basis for interpretation is critical and this is harder to discuss in general terms. Much more is typically required than the model provides a set of probability distributions yielding an adequate fit to the data under analysis.

Some desiderata for consideration are as follows:
1. the model should incorporate information from previous similar studies, allowing testing of consistency with these;
2. more specifically the model should use qualitative and quantitative ideas from subject-matter background and theory and allow testing consistency with these;
3. the model should be suggestive of a possible data generation process;
4. the model should contain parameters of interest that individually have clear substantive interpretation;
5. the "error structure" should allow realistic assessment of precision of the primary conclusions;
6. the "error structure" should take suitable account of any peculiarities of the data collection process;
7. the model should be consistent with the data in key respects.

3. Specificity of Objectives

An important general issue arises when the objective is a specific one of relatively short-term forecasting of one or more quantities. Is it better to use some simple quite empirical technique, disregarding some of the more ambitious desiderata listed in Section 2, or should a deeper analysis be attempted, based in some sense on understanding the system under study? Empirical evidence on forecasting performance, for example from the so-called forecasting competitions (Fildes & Makridakis, 1995) is ambivalent. Presumably the longer the time horizon the greater the preference for the second kind of approach.

At a more technical statistical level it is common to distinguish between estimation of unknown parameters and prediction of as yet unobserved random variables, the latter being relatively understudied. In Bayesian formulations the distinction at one level disappears, all unknowns being treated as random variables. At another level, in terms of the kind of model formulation to be adopted explicitly or implicitly, the broader issue remains.

4. Modes of Inference

Over the last 100 years and more there has been much discussion of the nature of probability and the role of Bayes's theorem in statistical inference. Bayesian approaches are of three very different kinds: the entirely uncontroversial use of empirical Bayes methods to represent a large number of essentially similar parameters via their frequency distribution, secondly by the use of standardized reference priors, and thirdly the explicitly personalistic approach.

Implementation of these requires integration often in a large number of dimensions. A major technical advance recently has been the implementation of Markov Chain Monte Carlo (MC-squared) methods introduced originally in physics. For some applications in the context of financial time series, see Shephard (1996). The methods are applicable also to various non-Bayesian likelihood based problems.

The use of reference priors goes back at least to Laplace and more recently to the systematic account by Jeffreys (1961), whose work derives to some extent from Keynes's doctoral thesis. For recent developments, see Bernardo (1979) and Berger and Bernardo (1992). For many relatively standard estimation procedures the differences from neo-Fisherian likelihood-based inference are fairly minor. Recent research in this has concentrated on higher-order asymptotic theory and the study of modified likelihood functions (Barndorff-Nielsen & Cox, 1994). There seems to be scope for fruitful investigation of the relation between higher-order asymptotics, Markov Chain Monte Carlo, and the so-called bootstrap and on the implications of all these for more complex econometric techniques such as cointegration. The issues can be critical whenever the number of nuisance parameters is relatively large.

The explicit introduction of additional information via a personalistic prior density raises much more challenging conceptual considerations and cannot be discussed adequately here.

One general comment is that issues about mode of inference within a given formulation may often be less critical than the choice of model itself.

5. Statistical Decision Theory

The previous discussion in effect presupposes that the object of analysis is the interpretation of data, usually attaching some indication of precision to estimates and predictions. There is another strand to statistical thinking that emphasizes decision making. The essence was set out in a remarkable early paper by Neyman and Pearson (1933), motivated probably in part by industrial inspection problems. The book by Wald (1950) was highly influential for a period leading to claims in theoretical circles that all formal statistical issues should be viewed in decision-theoretic terms. While vestiges of this view remain, and the notion of clarifying the purposes of analysis is clearly important, many statisticians probably see statistical decision theory as a relatively small part of their subject.

Wald, who had a strong interest in economics, supposed that utilities are clearly defined but prior probability enters his analysis only as a technical device for producing complete classes of admissible decision functions. There is nowadays a fairly broad agreement that the separate problems in determining prior probabilities and utilities are related and similar and that a satisfactory quantitative treatment requires a fully Bayesian approach.

Perhaps the main difference between the treatments of decision problems in the economic and in the statistical literatures is that mathematical economists, although perhaps not econometricians, seem to relate utility strongly, if not exclusively, to money, often in effect if not in principle linearly, whereas statistical discussions do so less, thereby to some extent reducing the concept of utility to circularity.

6. Interpretation of Observational Studies

In a sense, a number of the desiderata just raised aim to mitigate the limitations on the interpretation of observational as opposed to experimental studies. Often this centers on the question: what is the substantive meaning of a regression coefficient (using regression in a very broad sense as concerned with the dependence of one or more response variables on one or more explanatory variables)?

Two perhaps rather extreme examples of overinterpretation of regression coefficients are the conclusion that each implementation of the death penalty in the U.S. saved about seven lives and, secondly, inferences from a time series study of tobacco consumption, prices, and advertising expenditure concerning the likely consequences of making tobacco advertising illegal.

While the mathematical interpretation of a regression coefficient is clear as the statistical analogue of a partial derivative, the difficulties in interpreting regression coefficients in terms of the effect of interventions in the system stem from a number of sources:
1. the "laws" governing the system may change, especially under large interventions, cf. the Lucas critique;
2. explanatory variables omitted from the regression system, and perhaps not measured, or even whose existence is not appreciated, may not

change under intervention in the way that they have changed in the data under analysis;
3. other explanatory variables included in the regression equation, and therefore implicitly held fixed when the intervention variable changes, should themselves have imposed changes whose form may not be easily appreciated.

The three points are, of course, interrelated. The last is in a sense a technical error and can be largely corrected, but the first two points are more worrying.

One approach to the clarification of complex relations is the use of graphical representations, stemming from the geneticist Sewell Wright's notion of path analysis, introduced into econometrics by H.O.A. Wold. Recent work on this has moved away from the notion of decomposing dependencies into components, via generalizations of the formula relating total to partial regression coefficients,

$$\beta_{Y1} = \beta_{Y1.2} + \beta_{Y2.1} \beta_{21}$$

where Y is regressed on two explanatory variables X_1, X_2 and, for example, in Yule's notation, β_{Y1} is the total regression coefficient of Y on X_1 and $\beta_{Y1.2}$ is the corresponding partial regression coefficient given X_2. This is a statistical generalization of the elementary calculus formula relating ordinary and partial derivatives.

Rather, the emphasis is now on representing relatively complex systems of conditional independencies either in the context of probabilistic expert systems (Spiegelhalter et al, 1993) or for analysis of empirical data (Cox & Wermuth, 1996).

The interpretation of regression coefficients is often connected with issues of causal interpretation. Statisticians have tended to take a very cautious position over this, especially as regards observational studies, some taking the line that causal inference is possible only from randomized experiments. (Randomized experiments have, in my judgment, made a massive contribution to human welfare in connection with clinical trials but they too can raise major problems of interpretation). To a limited extent differences of opinion over causality are issues of definition. Experience does, however, suggest that the traditional view in epidemiology that risk factors and causal agents should be firmly distinguished is wise.

Bradford Hill (1965) gave eight conditions under which a causal interpretation of observational studies becomes more plausible. King, Keohane and Verba (1984) have discussed these from a political science viewpoint; see also Cox and Wermuth (1996, pp. 219-226). It would be interesting to have a parallel econometric analysis. By contrast there is coming from the artificial intelligence literature a much more positive attitude (Pearl, 1995; Spirtes, Glymour and Scheines, 1993). These include the development of computer programs that will determine causality even from single cross-sectional studies: Despite the considerable interest of these developments, the use of the word causality with its normal scientific connotation of some understanding of underlying process seems unwise. Similar remarks apply to the time series notion of Granger-causality; the special structures of conditional independence encapsulated in that notion can be of much interest but ambiguities of interpretation preclude the immediate inference of causality in any deep sense.

One of Bradford Hill's conditions concerned the possibility of drawing stronger inferences from so-called natural experiments. In his context these were typically massive interventions, such as natural or man-produced disasters, with such strong effects that anomalous behavior following the intervention can safely be ascribed to that intervention and not to unobserved confounders. In econometrics the term is used more widely to cover situations in which, even though the effects involved may be small, there are reasonable grounds for assuming that something approaching randomization has taken place. This is typically expressed via plausible assumptions about instrumental variables. An interesting example is the careful analysis by Angrist and Krueger (1991) via highly aggregated data of the effect of duration of schooling on earnings in which the instrumental variable month of birth has an association both with duration of schooling and earnings. It is argued that the association is evidence that duration of schooling is a suitable instrumental variable.

Rosenbaum (1995) has given a careful account of the general problems of design, analysis, and interpretation arising with observational studies.

7. Testing Goodness of Fit

Models are by their nature idealized and it is unreasonable to expect them to describe all aspects of the data. In very empirical approaches to statistical analysis consistency with the data is often regarded as of overwhelming importance but it seems doubtful if this is wise. Tests can be regarded as ones of
1. the adequacy of the underlying theory or formulation, failure to fit the data indicating a need to rethink the whole basis of analysis;
2. details of formulation, such as number of lags needed, failure indicating a need for relatively minor modification;
3. important aspects of error structure, affecting in a critical way the assessment of precision and the efficiency of estimation;
4. relatively minor aspects of error structure.

A difficulty common in some fields of application is the clash between statistical significance and substantive significance and in particular the possibility that an effect may be highly significant statistically but yet too small to be of substantive importance. With the possible exception of the analysis of some kinds of financial data, this difficulty seems likely to be uncommon in econometrics, where the models fitted are often somewhat elaborate relative to the amount of information in the data, so that any effect found statistically significant is likely to be of subject matter concern.

A general strategy is that all significant lack of fit is to be reported and any argument that the discrepancy is substantively unimportant explicitly stated and as far as feasible justified. The obligation to report any lack of fit found is protection against extreme arbitrariness in model choice.

Issues of goodness of fit are connected with what in the econometric literature, but not elsewhere, is called *calibration*. With complex systems it may be helpful to divide the response features, i.e., those which are not to be held fixed in the model, into those of direct interest, those which are of indirect interest, and those which are unimportant or which are accepted as poorly represented in the proposed model. The first set are to be used in fitting, for example by some form of the generalized method of moments, and in testing goodness of fit. Appreciable lack of fit will not be acceptable. If the second set of features, not used in fitting, are well represented the model gains in general credibility. The fit of the third set of features is not expected to be good and this is not regarded as important. Of course the division depends totally on context.

These remarks are of most relevance for fairly complicated problems. A relatively simple example would be the fitting of a continuous time diffusion-like model to share prices. The first set of features would be the mean and covariance structure of returns over time spans of say 1 hour, and longer. The second set of features might concern the extremes of the process. The third might be very short-term properties of the process on a time scale where individual transactions are relevant, calling for a different type of continuous time model in which the point process structure is recognized.

8. Sensitivity Analysis

Large computer models are now quite widely used in many fields certainly including economics and a number are mentioned in companion papers. Their use may be either to gain insight by exploring the behavior of complex systems under idealized conditions or to develop predictions for policy decisions, typically by estimating the consequences of intervention. It is easy to feel considerably more comfortable with the former than with the latter, certainly if the latter is not preceded by the former!

By contrast with the analytical solution of simple models, whose value for developing qualitative understanding can be great, the computer models require the assignment of numerical values to a large number of unknown parameters (coefficients). Some numerical values may be derived from explicit statistical analysis of relevant data, others may be chosen based on relatively informal "historical" analysis while yet others may be assigned via informal expert judgment, which, if not derived from well-formulated evidence, can be particularly suspect.

In all cases, whether the model is deterministic or stochastic, study of the reliability of the conclusions to the numerical assumptions made, is highly desirable, indeed should be mandatory.

This raises some technical statistical issues in particular regarding the model as a multidimensional system to be explored by highly fractionated design (Sachs et al. 1989; Welch et al. 1992). It is an open issue as to how far dependence on assumptions is best explored in this way by designed sensitivity analysis and how far by so-called elicitation of priors for the unknowns, leading to a posterior distribution for the quantities of interest giving a composite measure of uncertainty. It seems to me that some element at least of the former is desirable.

9. Panel Data

A very interesting field in which there are parallel and largely complementary literatures (emanating from the econometric, the sociological, the biostatistical, and the industrial reliability fields), is the analysis of panel data taken broadly to mean the analysis of many short time series. Technically the statistical properties come from consideration of a large number of supposedly independent individuals rather than from the reliance on the ergodic properties of a single long series.

A special kind of panel data arises with data defined by point events such as the beginning and ending of periods of employment. There is a vast predominantly biostatistical literature on survival data, the situation where there is essentially a single interval under analysis, the focus being its dependence on explanatory variables; see, for example, Cox and Oakes (1984) and from a viewpoint emphasizing the connection with martingales (Andersen et al. 1993). The sociological literature, under the name event-history analysis (Blossfeld, Hamerle & Mayer, 1992; Blossfeld & Rohwer, 1995), emphasizes the analysis of short series of events possibly of different types. There are strong connections also with the study of Markov and semi-Markov processes with discrete states; see, for example, Janssen (1986). Lancaster (1990) has given a very interesting account from an econometric perspective.

10. Conclusion

The literature on econometric and statistical techniques has a good deal of naturally arising overlap, with inevitably some substantial duplication and some differences of emphasis. Of more interest in many ways is the relation between the approaches to specific problems. The statistical ethos, superficially at least, puts more emphasis on problems of study design, on issues of measurement and data quality, on a desire for simple models, and on a purely empirical approach, and moreover is strongly influenced, indirectly at least, by an interest in the experimental as well as the observational sciences. The emphasis on assessment of precision tends to lead to a cautious, sometimes overcautious, attitude to the interpretation of data. The econometric tradition appears to involve more willingness to fit rather complex models to limited data. The present chapter touches on just a few of the issues involved.

If there is a unifying theme to late 20th century applied mathematics it is *nonlinearity*. With the exception of the special models developed to study volatility, most of the models discussed in the very applied econometric literature appear to be linear and it is perhaps in moving away from linear models that the most scope for methodological development lies.

I am very grateful to Dr Neil Shephard for helpful suggestions.

11. References

Andersen P.K., Borgan, O., Gill, R.D., & Keiding, N. (1993). *Statistical models based on counting processes*. New York: Springer.
Angrist, J.D. and Krueger, A.B. (1991). Does compulsory school attendance effect schooling and earnings? *Q.J. Economics*, 106, 979-1014.
Barndorff-Nielsen, O.E. and Cox, D.R. (1994). *Inference and asymptotics*. London: Chapman and Hall.
Berger, J.O. and Bernardo, J.M. (1992). On the development of the reference prior method (with discussion). *Bayesian Statistics* 4, 35-60. (J.M. Bernardo, J.O. Berger, A.P. Dawid and A.F.M. Smith, eds). Oxford University Press.
Bernardo, J.M. (1979). Reference prior distributions for Bayesian inference (with discussion). *J.R.Statist. Soc.*, B 41, 113-147.
Blossfeld, H.-P., Hamerle, A. and Mayer, K.U. (1992). *Event history analysis: statistical theory and application in the social sciences*. Hillsdale, NJ: Lawrence Erlbaum Associates.
Blossfeld, H-P. and Rohwer, G. (1995). *Techniques of event history modeling*. Hillsdale, NJ: Lawrence Erlbaum Associates.
Bradford Hill, A. (1965). The environment and disease: association or causation? *Proc. Roy. Soc. Medicine*, 58, 295-300.
Cox, D.R. and Wermuth, N. (1996). *Multivariate dependencies*. London: Chapman and Hall.
Cox, D.R. and Oakes, D. (1984). *Analysis of survival data*. London: Chapman and Hall.
Fildes, R. and Makridakis, S. (1995). The impact of empirical accuracy studies on time series analysis and forecasting. *Int. Statist. Rev.*, 63, 289-308.
Janssen, J. (editor) (1986). *Semi-Markov models: theory and application*. New York: Plenum.
Jeffreys, H. (1961). *Theory of probability*. Clarendon Press: Oxford.
King, G., Keohane, R.O. and Verba, S. (1994). *Designing social enquiry*. Princeton University Press.
Lancaster, T. (1990). *The econometric analysis of transition data*. Cambridge University Press.
Neyman, J. and Pearson, E.S. (1933). The testing of hypotheses in relation to probabilities a priori. *Proc. Camb. Phil. Soc.*, 24, 492-510.
Pearl, J. (1995). Causal diagrams for empirical research (with discussion). *Biometrika*, 82, 669-710.
Rosenbaum, P.R. (1995). *Observational Studies*. New York: Springer.
Shephard, N. (1996). Statistical aspects of ARCH and stochastic volatility. In *Time series models*, D.R. Cox, D.V. Hinkley and O.E. Barndorff-Nielsen (editors). London: Chapman and Hall.
Sachs, J., Welch, W.J., Mitchell, T.J. and Wynn, H.P. (1989). Design and analysis of computer experiments (with discussion). *Statistical Science*, 4, 409-435.
Spiegelhalter, D.J., Dawid, A.P., Lauritzen, S.L., and Cowell, R.G. (1993). Bayesian analysis in expert systems (with discussion). *Statistical Science*, 8, 219-283.
Spirtes, Glymour & Scheines, 1993.
Wald, A. (1950). *Statistical decision functions*. New York:Wiley.
Welch, W.J., Buck, R.J., Sacks, J., Wynn, H.P., Mitchell, T.J. and Morris, M.D. (1992). Screening, predicting and computer experiments. *Technometrics*, 34, 15-26.

Received: July 1999

4
THE STATISTICAL CENTURY

Bradley Efron
Department of Statistics
Stanford University

My chosen title, "The Statistical Century," is about as general as one can get for a statistics talk. It is also ambiguous. In fact it refers to two centuries: what has happened to statistics in the 20th century, and what might happen to us in the 21st.

Of course there's a close connection between the two centuries. The past is an uncertain guide to the future, but it's the only guide we have, so that's where I'll begin.

There is some peril lurking here. Predictions are dangerous where future readers get to grade the speaker for accuracy. But I'll be happy with just a partial success, getting across the flavor of the statistical history.

My first point, and perhaps the most important one of the whole chapter, is to remind you that statisticians are heirs to a powerful intellectual tradition— a few centuries old, but with by far the greatest successes, both in terms of achievements and influence, since 1900.

Viewed as a whole, the 20th century was decisively successful for the rise of statistical thinking. It's become the interpretive technology of choice in dozens of fields: economics, psychology, education, medicine, and lately in the hard sciences too, geology, astronomy, and even physics.

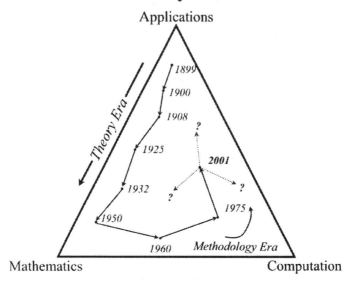

Fig.4.1 Twentieth century statistics

The diagram (Fig.4.1) is a little automobile trip through the statistical century, travelling between the three poles of statistical development: applications, mathematics, and computation. Autos are the appropriate vehicle, themselves being a 20th century phenomenon.

Our trip begins in 1899, just before the start of the 20th century. There was already an impressive list of statistical techniques available: Bayes theorem, Least Squares, the Central Limit theorem, Galton Correlation, and Regression, Binomial and Poisson methods.

But the list existed as a collection of useful adhoc methods, not as a coherent theory. There was no real field called "Statistics," just some methodology clustered around particular application areas, like economics, public health, and government. That's why I've started the trip at the "applications pole." I doubt most scientists would have thought that there could be an independent field of statistics. Astronomers have stars, there's rocks for geologists, but what would be the subject matter of statistics? Information science was an idea that hadn't been formulated yet. It was cooking though, particularly in the mind of Karl Pearson.

1900: A bold first step into the new century. Pearson's chi-squared paper was a qualitative leap in applying powerful new mathematics (matrix theory) to statistical reasoning. It greatly raised the level of mathematical sophistication in statistics. The next year, 1901, Pearson founded *Biometrika*, the first modern statistics journal. (The first issue has a charming essay by Galton, including a characteristically clever explantion of CDFs.)

1908: The student's t statistic. This had a particularly strong influence on Fisher's ideas pointing toward a deeper understanding of statistical inference, and one that applied in small samples, not just to census-level data. Student's paper greatly raised our level of inferential sophistication.

1925: Fisher's great estimation paper. It initiated a host of fundamental ideas: consistency, sufficiency, efficiency, Fisher information, Maximum likelihood estimation. Its biggest contribution was Optimality, the Best one can do in a given estimation problem. Optimality is the mark of maturity for a scientific discipline (missing, and badly needed in Computer Science). I mark 1925 as the year statistics went from a collection of ingenious techniques to a coherent discipline.

Fisher had some reservations about optimality, which I'll soon mention, but once the mathematical optimality genie was out of the bottle, our automobile trip turned into a joyride toward the mathematics corner of the triangle.

1932 represents Neyman and Pearson's classic paper, optimality theory for testing problems. It seems this should have pleased Fisher but instead it initiated a life-long feud with Neyman. Some of this was personal jealousy, but not all: Fisher felt that Neyman's approach could be Optimal without being Correct. For instance, given a random sample from a Cauchy distribution, the optimal [that is the Shortest] confidence interval for the distribution's center was Incorrect: one should actually use intervals of varying lengths, depending on the amount of Fisher information in the given sample. Correctness is one of Fisher's ideas that is definitely out of favor right now, but, as I'll mention later, it is hovering in the wings of some current developments.

1950 stands for Wald's decision theory and also Savage & DiFinniti's subjective Bayes formulation.

By *1960*, when I began my statistics education, jumping on to the auto, my timing was poor: it looked like our auto trip was going to end with a disasterous crash into the math corner, which would also be the end of statistics as an independent field. It's no coincidence that the 1960s marked the Nadir of the statistic profession's influence on statistical applications. Scientists can do statistics without a statistics profession, just not very well.

Just in time though the influence of electronic computation began to be felt. Tukey's influential paper on the Future of Data Analysis deserves some of the credit. The auto swerved suddenly toward the computation corner.

Fortunately we didn't turn into a branch of Computer Science either. A further swerve took us back toward applications, but now with a statistics profession better armed with mathematical and computational equipment. A very happy development: we've became a lot more useful to our fellow scientists, the ones who need to use statistical reasoning, our clients in other words.

I've located *2001* at the center of the triangle, a healthy place to be. The question marks indicate that I really don't know which direction(s) we'll be going next. Some speculation will follow, but there's one more feature of the triangle that will help me frame the question.

I've labelled *1900-1950* the "Theory Era." That's when most of our current storehouse of fundamental ideas was developed. It was the golden age of statistical theory, the time when the Inside of our field was cultivated.

1960-2000 is labelled as the "Methodology Era." The theory from the golden age was harnessed with the power of modern computation to make statistics enormously useful to our scientific customers. This has been a golden age for developing the "Outside" of our field.

The next figure (Fig.4.2) is a list of 12 important postwar developments that have had a big effect on applications. The list isn't exhaustive, just my personal favorites, along with the convenience of being able to list them in pairs. I've restricted myself to 1950-1980, allowing 20 years for a little bit of perspective.

---

*Nonparametric & Robust methods

*Kaplan-Meier & Proportional Hazards

*Logistic Regression & GLM

*Jacknife & Bootstrap

*EM & MCMC

*Empirical Bayes & James-Stein estimation

Fig.4.2 Twelve postwar developments.

1. Nonparametric and Robust Methods

Wilcoxon's little paper set off an avalanche of new theory and methodology, as well as giving us the most widely-used post-war statistical tool. Robust methods, a la Box, Tukey, and Huber, have had a slower growth curve, but have made a marked impression on the consciousness of statistical consumers. Robust and nonparametric regression methods ("smoothers") seem to be of particular current interest.

2. Kaplan-Meier and Proportional Hazards

Progress in survival analysis has had a spectacular effect on the practice of biostatistics. From my own experience, with half of my appointment in the

Stanford Medical School, the practice of biostatistics is much different, and better, now than in 1960.

Stigler's 1994 Statistical Science paper on which papers and journals have had the greatest influence, puts Kaplan-Meier and Cox's proportional hazards paper as numbers 2 and 3 respectively on the postwar citation list, following only Wilcoxon.

3. Logistic Regression and GLM

The extension of normal-theory linear models to general exponential families has had its biggest effect on binomial response models, where logistic regression has replaced the older probit approach. (Again, to a huge effect on biostatistics). Generalized linear models are now more popular for dealing with heteroskadastic situations than the Box-Cox transformation approach, highly popular in the 1960s. Overall there is now a much more aggressive use of regression models for non-Gaussian responses.

4. Jacknife and Bootstrap

Quenouille's original goal of nonparametric bias estimation was extended to standard errors by Tukey. The bootstrap did more than automate the calculation of standard errors and confidence intervals: it automated their theory. The success of resampling methods has encouraged attempts to automate and extend other traditional areas of statistical interest, in particular Bayesian inference. This brings us to the next pair,

5. EM and MCMC

These are computer-based techniques for finding maximum likelihood and Bayes estimators in complicated situations. MCMC, including the Gibbs sampler, has led to a revival of interest in the practical applications of Bayesian analysis. Personally I've become much more interested in Bayesian statistics in its new more realistic, less theological rebirth.

6. Empirical Bayes and James-Stein Estimation

This closely related pair of ideas is the biggest theoretical success on my list, and the biggest underachiever in volume of practical applications. The irony here is that these methods offer potentially the greatest practical gains over their predecessors. Wilcoxon's test isn't much of an improvement over a t-test, at least not in experienced hands, but in favorable circumstances an empirical Bayes analysis can easily reduce errors by 50%. I'm going to say more about this paradox soon.

This list is heavily weighted toward the outside development of statistics, ideas that make statistics more useful to other scientists. It's crucial for a field like statistics that doesn't have a clearly defined subject area (such as rocks or stars) to have both inside and outside development. Too much inside and you wind up talking only to yourself. The history of academic mathematics in the 20th century is a cautionary tale in that regard.

However outside by itself leaves a field hollow, and dangerously vulnerable to hostile takeovers. If we don't want to be taken over by Neural Networkers or Machine Learners or Computer Scientists, people who work exclusively in some interesting area of applications, and have a natural advantage of their own turf, we have to keep thinking of good statistical ideas, as well as making them friendly to the users.

Nobody was ever better at both the inside and outside of statistics than Fisher, nor better at linking them together. His theoretical structures connected seamlessly to important applications (and as a matter of fact, caused those applications to increase dramatically in dozens of fields).

I wanted to draw attention to a few salient features of Fisherian thinking— features which have been an important part of past statistical development, and by implication have to be strong candidates for future work.

Bayes/Frequentist Compromise:

When Fisher began his work in 1912, Bayesian thinking of the "put a uniform prior on everything" type advocated by Laplace was still prominant. Fisher, dissatisfied with Bayes, was not a pure frequentist (though his work helped launch the frequentist bandwagon). This was more a question of psychology than philosophy. Bayesian statistics is an optimistic theory that believes that one has to do well only against the "Correct" prior (which you can know.) Frequentists tend to be much more conservative, one definition of frequentism is *to do well*, or at least not disasterously, against Every possible prior. Fisher fell somewhere in the middle of the optimist-pessimist statistical spectrum.

To my reading Fisher's work seems like a series of very shrewd compromises between Bayes and frequentist thought, - but with several unique features that nicely accommodate statistical philosophy to statistical practice. I've listed some main points in the following table.

4. THE STATISTICAL CENTURY

- Bayes / Frequentistic compromise but...

- Direct interpretation of likelihood (MLE)

- Reduction to simple cases (sufficiency, Ancillarity, Conditioning ...)

- Automatic algorithms ("Plug-in")

- Emphasis on (nearly) unbiased estimates and related tests.

Fig.4.3 Fisherian Statistics.

Direct Interpretation of Likelihoods

This is the philosopher's stone of Fisherian statistics. If a complete theory existed then we could shortcut both Bayes and frequentist methodology and go directly to an answer that combines the advantages of both. Fisher's big success here was Maximum Likelihood Estimation, which remains the key connector between statistical theory and applications. Could there possibly be a more useful idea? Maybe not, but it has a dark side that we'll talk about soon.

Fisher had a sensational failure to go with the MLE's success: Fiducial Inference, generally considered just plain wrong these days. However Fisher's failures are more interesting than most people's successes, and the goal of the Fiducial theory, to give what might be called Objective Bayesian conclusions without the need for Subjective priors, continues to attract interest.

Reduction to Simple Cases

Fisher was astoundingly resourceful at reducing complicated problems to simple ones, and thereby divining "correct" answers: Sufficiency, Ancillarity, Conditional arguments, transformations, pivotal methods, asymptotic optimality.

Only one major reduction principle has been added to Fisher's list, "invariance," and it's not in good repute these days. We could really use some more reduction principles these days, in dealing with problems like model selection, prediction, or classification, where Fisher hasn't done all the hard work for us.

Automatic Algorithms

Fisher seemed to think naturally in algorithmic terms: Maximum Likelihood Estimation, Analysis of Variance, Permutation Tests, are all based on computational algorithms that are easy to apply to an enormous variety of situations. This isn't true of most frequentist theories, say minimax or UMVU estimation, nor of Bayesian statistics either, unless you're a Laplacianist — not even MCMC will tell you the Correct prior to use. "Plug-In," an awkward term but I don't know a better one, refers to an extremely convenient operational principle: to estimate the variability of an MLE, first derive an approximate formula for the variance (from Fisher Information considerations) and then substitute MLE's for any unknown parameters in the approximation. This principle, when it can be trusted, gives the statistician seven-league boots for striding through practical estimation problems.

The bootstrap is a good modern example of the plug-in principle in action. It skips the hard work of deriving a formula, by using the computer to directly plug in the observed data, for the calculation of a variance or confidence interval.

Exactly when the plug-in principle can be trusted is a likely area of future basic research. The principle is deeply connected with unbiased estimation, and its theory gets dicey when there's too many parameters to plug in.

I've gone on a little about Fisher because he is so crucial to 20th century statistics, but also because we seem to be reaching the limits of where his ideas, and those of Student, Neyman, Pearson, Savage, De Finetti, can take us.

Fisherian statistics has been criticized, justly, for its over-reliance on Normal-theory methods. Developments like generalized linear models, nonparametrics, and robust regression have helped correct this deficit.

We haven't been nearly as successful in freeing ourselves from another limitation, an excessive dependence on unbiasedness. It wouldn't be misleading to label the 20th century as "100 years of unbiasedness." Following Fisher's lead, most of our current statistical practice revolves around unbiased or nearly unbiased estimates (particularly MLEs) and tests based on such estimates. It is the power of this theory that has made statistics so important in so many diverse fields of inquiry, but as we say in California these days, it is power purchased at a price. (That's provincial humor; we ran short of electricity last summer, and saw a nasty increase in our bills).

The price of the classic approach is that the theory only works well in a restricted class of favorable situations. "Good experimental design" amounts to enforcing those conditions, basically a small ratio of unknown parameters to observed data points. As a matter of fact the ideal is to isolate the question of interest in a Single crucial parameter, the "Treatment Effect," only achievable, and at great expense and effort, in a randomized double-blinded controlled experiment.

This is a powerful idea but a limited one and we seem to be rapidly reaching the limits of 20th century statistical theory. Now it's the 21st century and we are being asked to face problems that never heard of good experimental design. Sample sizes have swollen alarmingly, and the number of parameters more so.

Even a moderate sized medical data base can contain millions of data points referring to 100,000 questions of potential interest.

The response to this situation, which has come more from those who face such problems than from the professional statistics community, has been a burst of adventurous new prediction algorithms: neural networks, machine learning, support vector machines, all under the generic term "data mining." The new algorithms appear in the form of black boxes with enormous numbers of knobs to play with, that is with an enormous number of adjustable parameters. They have some real successes to their credit, but not a theory to explain those successes or warn of possible failures.

The trouble is that the new algorithms apply to situations where biased estimation is essential, and we have so little biased estimation theory to fall back upon. Fisher's optimality theory, in particular the Fisher Information Bound, is of no help at all in dealing with heavily biased methodology. In some sense we are back at the beginning of our automobile trip, with lots of ingenious adhoc algorithms, but no systematic framework for comparing them.

I've been speaking in generalities, which goes with the turf in speaking about a broad topic like statistical history. However, I wanted to discuss a specific "big-data" problem, of the type in which classical theory is inadequate. I've chosen a microarray problem, mainly because I'm working on it right now, with Rob Tibshirani and John Storey, but also because it is nicer than most such problems in at least giving a hint how it might be approached theoretically.

Microarrays measure "expression levels," how active a particular gene is in the workings of a living cell— for example is gene x more active in a tumor cell than in a normal cell? Expression levels have been measurable for some time, but it used to take on the order of a day to measure just one gene. The charm of microarrays is that they can measure thousands of expression levels at the same time. This represents a thousand fold advance for biologists, the same kind of lift statisticians got from the introduction of computers. Along with all this data comes some serious data analysis problems, and that's where statisticians enter the picture.

Table 4.1 BRCA1, BRCA2 Microarray Experiment. 3200 Genes, 15 Microarrays.

Which Genes Express Differently?

	---BRCA1---	---BRCA2---	W	p-value
	Mic1...mic7	mic8...mic15		
Gene 1	-1.29 . -0.57	-0.70.. 0.16	83	.025
Gene 2	3.16 .. 0.60	-1.08.. -0.28	45	.028
Gene 3	2.03 .. -0.78	0.23.. 0.28	50	.107
Gene 4	0.32 .. 1.38	0.53.. 2.83	64	.999
Gene 5	-1.31 .. 0.40	-0.24.. 1.15	81	.047
Gene 6	-0.66 .. -0.51	-0.41.. -0.10	67	.737
Gene 7	-0.83 .. 0.22	-0.99.. 0.43	58	.500
Gene 8	-0.82 .. -1.44	0.31.. 1.23	71	.430
Gene 9	-1.07 .. -0.34	-0.55.. -0.87	55	.308
Gene 10	1.19 .. 0.22	-1.21.. 0.58	58	.500
.	.	.	.	.
.	.	.	.	.
.	.	.	.	.
Gene 3200	. . . . . .	. . . . . .	84	.018

From Hedenfalk et al., February 2001, Gene-Expression Profiles in Hereditary Breast Cancer, *New England Journal of medicine*, 344; 539-548

In the experiment I'm going to discuss, from Hedenfalk et al. in the February 2001 issue of the *New England Journal of Medicine*, each microarray was reporting expression levels for 3200 individual genes (the same 3200), in breast cancer tumor cells. The experiment involved 15 microarrays, one for each of 15 tumors from 15 different cancer patients. The first 7 tumors came from women known to have an unfavorable genetic mutation called "BRCA1," while the last 8 came from women having another unfavorable mutation "BRCA2." The mutations, which are on different chromosomes, are both known to greatly increase the risk of breast cancer. A question of great interest was: which genes express themselves differently in BRCA1 as compared with BRCA2 tumors?

4. THE STATISTICAL CENTURY

This is a fairly modest-sized experiment by current microarray standards, but it produced an impressive amount of data: a 3200 by 15 data matrix, each of the 15 columns being the 3200 measured expression levels for one tumor.

If we had just the 15 expression numbers for a single gene, like gene 1 on Table 4.1, the first row of the big matrix, we might very well run a Wilcoxon two-sample test, the post-war favorite: rank the 15 numbers, and add up the BRCA2 ranks to get the Wilcoxon rank-sum statistic "W." Unusually big or unusually small values of W then indicate a significant difference between BRCA1 and BRCA2 expression for that gene.

For gene 1 the 8 BRCA2 measurements were mostly larger than the 7 BRCA1's, giving a big W, 83. The two-sided p-value for the usual Wilcoxon test is .024 so if we were only considering gene 1 we would usually reject the null hypothesis of no difference and say "yes, BRCA1 tumors are different from BRCA2's for gene 1." Gene 2 is significant in the other direction, showing greater expression in BRCA1's, while gene 3 is not significant in either direction.

The next figure (Fig.4.3) shows the histogram of all 3200 W scores, with a dashed curve called $\hat{f}(w)$ drawn through it.

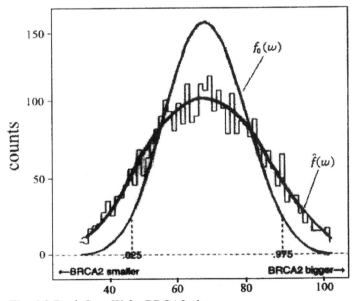

Fig. 4.3 Rank Sum W for BRCA2 plates.

We see that $\hat{f}(w)$ is a lot wider than $f_0(w)$, the Wilcoxon distribution (solid curve) that would apply if nothing was going on— that is if all 3200 genes satisfied the null hypothesis of no BRCA1-BRCA2 difference. Certainly this can't be the case. Six-hundred and fourteen of the 3200 W's have the null hypothesis of no difference rejected by the usual .05 two-sided test. But we'd expect 160 false rejections, 5% of 3200, even if nothing was happening. We seem to have a horrendous multiple comparison problem on our hands. How can we decide which of the 3200 genes behaved "Differently" in the two types of tumors, and which did not?

One answer, and there are others, is to look at the problem from an empirical Bayes point of view, a la Robbins and Stein, our underachiever pair on the list of breakthroughs.

A very simple Bayes model assumes that there are just two classes of genes: those that are expressed the same for BRCA1 and BRCA2, and those that express differently; and that W has the usual Wilcoxon distribution for the "Same" class, but some other distribution for genes in the "Different" class. We don't have a theory to say what the "Different" distribution is, but the $\hat{f}(w)$ curve provides strong empirical evidence for what its shape must be.

It turns out that it is easy to write down Bayes rule for the a posteriori probability that a gene is in the "different" class given its W score: as shown on the next graph, the rule depends only on the ratio of the two curves $f_0(w)$ and $\hat{f}(w)$.

A posteriori Probability of "Different":
$$\text{Prob}\{\text{Different}| W\} > 1 - f_0(w)/\hat{f}(w)$$

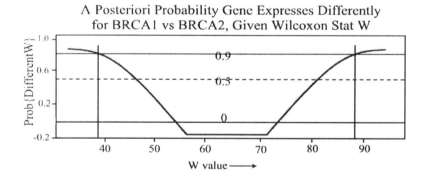

Fig.4.4 A Posteriori Probability Gene expresses differently for BRCA1 vs BRCA2, given Wilcoxon Stat W

The graph (Fig.4.4) shows the results for the tumor data: The heavy curve is the a posteriori probability of a gene being "different" as a function of its Wilcoxon score W. It turned out that in order to have the posterior probability of "different" greater than 90%, a gene's W score needed to be among the 3 smallest or 3 largest possible values.

Only 67 of the 3200 genes showed such strong results. This criterion is a lot more conservative than the naive one-at-a-time .05 tests, which gave 614 ".05 significant" genes, but it has good Bayesian justification, and also frequentist justification in terms of what is called the False Discovery Rate, a nice new simultaneous inference technique developed by Benjamini and Hochberg. I can't go into that here, except to say that it's usually a good sign when both Bayesian and frequentist reasoning point toward the same answer.

The main point of this example is that it is definitely post-Fisherian. It is empirical Bayes in the sense that it begins with a pure Bayes setup, but then estimates a crucial part of the resulting Bayes rule ($\hat{f}(w)$) from the data. Empirical Bayes is a different sort of Bayes/Frequentist compromise [to use I.J. Good's terminology] than Fisher envisioned.

I said that we have very little theory to fall back upon when we want to use biased estimation. The great exception is Bayesian theory, which provides a very satisfying framework for optimal decision-making, biased or not, when we have reliable prior information. Of course having a believable prior is the great practical impediment to Bayesian applications. The prior's absence in most scientific situations is what led Fisher, and most 20th century statisticians, away from Bayes.

Robbins' and Stein's crucial insight was that for problems with parallel structure, like the microarray situation, we can use the data itself to generate the prior.

From this point of view, having all those genes, 3200 of them, starts to look more helpful than distressing. For the purposes of learning about the prior, 3200 is preferable to say 10 or 20. Another way to say the same thing is that multiple comparison problems get easier when we have massive numbers of effects to compare, rather than just a few.

Most discussions about the future are really about the past and present, and this one is no exception, but here's one prediction that seems gold-plated to me: research on the combination of Bayesian and frequentist methodology will be a crucial part of 21st century statistics.

At my most hopeful I can imagine a new theory of statistical optimality emerging, one that says what's the Best one can do in our microarray problem, for example.

But maybe I'm being too hopeful here. The microarray example is atypical of large data sets in that its largeness derives from the parallel replication of individual small sub-experiments. This isn't the case if one goes data-mining through a large medical or supermarket data base.

On the other hand, statisticians are more than just passive processors of whatever data happens to come our way. Fisher's theory of efficient experimental design greatly influenced the form of 20th century data sets. Analysis of Variance fits an amazing number of situations, but that's at least partly because research scientists know that we can effectively analyse ANOVA data. If statisticians can demonstrate efficient ways of analyzing parallel data, then we'll start seeing more parallelism in data base design.

That's the end of the microarray example, except for a confession and an apology. When Carl Morris and I were doing Empirical Bayes work in the 1970s, we concentrated on Stein's parametric version of the theory, because it applied to much smaller data sets than Robbins' nonparametric methods. I must admit to thinking, and even saying, that Robbins' methods would never be of much practical use for "realistic" sample sizes. Microarrays prove that the meaning of realistic is changing quickly these days.

Someone once told me that the intersection of all the countries that have been called Poland is Warsaw. I mention this because if there is a geographical analogue to the statistics profession it might be Poland: strategically located, surrounded by powerful and sometimes aggressive neighbors, and still maintaining a unique and vital national heritage. Statistics had a much better 20th century than Poland, but Poland is still prominent on the map of Europe, and doing pretty well these days.

Statistics is doing pretty well these days too. We tend to think of statistics as a small profession, perhaps comparing ourselves with biology or computer science, but in fact we aren't so small in sheer numbers, or, more emphatically, in our importance to the general scientific enterprise (bigger than astronomy, a little smaller in numbers than geology).

How will the statistics profession do in the 21st century? That's up to you, particularly the younger members of the profession, who form the next wave of statistical researchers. Fields like ours prosper in proportion to their intellectual vitality. My list of 12 methodological advances spanned 1950-80, allowing myself 20 years hindsight for perspective. I have hopes that in, say, 2025 you'll be able to double the list, perhaps including some advances in basic theory, the inside of the profession, as well as useful methodology.

One can make a good case that we are the ones living in the golden age of statistics— the time when computation has become faster, cheaper, and easier by factors of a million, and when massive data collection has become the rule rather than the exception in most areas of science. Statistics faces many great intellectual challenges, not to mention the usual financial, societal, and academic ones, but our record in the 20th century should give us confidence in facing the 21st.

I thought I'd finish as I began, with a statistical triangle. This time the corners are our three competing philosophies, Bayesian, frequentist, and in the middle Fishersian, and I've tried to allocate the influence of the three philosophies on some important current research topics.

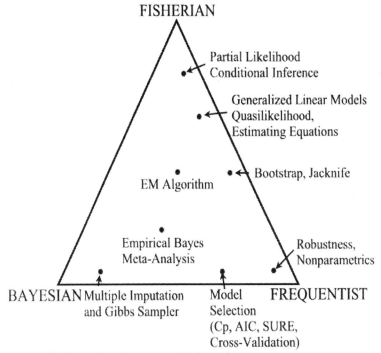

Fig.4.5 The Statistical Competing Philosophies

Some of the topics were easy to place: Partial Likelihood and Conditional Inference lie at the Fisherian pole, reflecting David Cox's position as our leading Fisherian. Robustness and Nonparametrics follow Huber and Tukey's preference for frequentism. The Bayesian pole obviously merits multiple imputation and the Gibbs sampler. I've placed the bootstrap and the jackknife, accurately I think, half-way between Frequentism and Fisherian thinking, and similarly Empirical Bayes halfway between the Frequentist and Bayesian corners.

Other topics aren't so clearcut, as you can see by my weasly placement of EM near the center of the figure. You might enjoy arguing with this picture, or placing your own favorite topics in the triangle.

One important conclusion from the diagram: all three of the competing statistical philosophies are alive and contributing to active research. Personally I've been happy to use whatever philosophy seems most effective for any given problem, but it's useful to at least know the intellectual tools that are available to us, and what their past successes have been.

Some recent developments in algorithmic prediction models don't seem to fit into the triangle very well. The "prediction culture," as Leo Breiman calls it, which includes neural networks, data-mining, machine learning, boosting, bagging, and support vector machines, has grown up in the front lines of the struggle with massive data sets. It is reasonable to expect ambitious new statistical enterprises to begin with a burst of pure data analysis. The innovators may even be openly scornful of the old statistical guard.

This is the teenage rebellion phase of new methodology. The great achievement of 20th century statistics was to put a firm logical foundation under older adhoc methods like regression and design; that is, to locate them inside the statistical triangle. This was a great aid to the development of better methodologies, and also to the development of a more powerful statistical theory.

The new generation of statistical researchers faces the same challenge with the prediction culture. At some point, the questions of optimality and inferential correctness have to be faced, even by the prediction culture.

Based on our past record, we can reasonably expect success. Maybe we can even hope for an expansion of the theoretical basis of statistical inference, so another chapter in 2025 will need to replace my triangle with a square.

We can all be proud of belonging to a profession that has done so much to foster clear thinking in science, government, industry, and society in general.

Received: April 2002, Revised: May 2002

5
FROM ASSOCIATION TO CAUSATION: SOME REMARKS ON THE HISTORY OF STATISTICS

David Freedman
Statistics Department
University of California, Berkeley

1. Introduction

In this chapter, I will look at some examples from the history of statistics—examples which help to define problems of causal inference from non-experimental data. By comparing the successes with the failures, we may learn something about the causes of both; this is a primitive study design, but one that has provided useful clues to many investigators since Mill (1843). I will discuss the classical research of Pierre Louis (1835) on pneumonia, and summarize the work of John Snow (1855) on cholera. Modern epidemiology has come to rely more heavily on statistical models, which seem to have spread from the physical to the social sciences and then to epidemiology. The modeling approach was quite successful in the physical sciences, but has been less so in the other domains, for reasons that will be suggested in Sections 4–7.

Regression models are now widely used to control for the effects of confounding variables, an early paper being Yule (1899); that is the topic of Section 4. Then some contemporary examples will be mentioned, including studies on asbestos in drinking water (Section 5), health effects of electromagnetic fields, air pollution, the leukemia cluster at Sellafield, and cervical cancer (Section 7). Section 8 discusses one of the great triumphs of the epidemiologic method-identifying the health effects of smoking. Other points of view on modeling are briefly noted in Section 9. Finally, there is a summary with conclusions.

2. La Méthode Numérique

In 1835, Pierre Louis published his classic study on the efficacy of the standard treatments for pneumonia: *Recherches sur les effets de la saignée dans quelques maladies inflammatoires: et sur l'action de l'émétique et des vésicatoires dans la pneumonie.* Louis was a physician in Paris. In brief, he concluded that bleeding the patient was a good treatment for pneumonia, although less effective than commonly thought:

> "Que la saignée a une heureuse influence sur la marche de la pneumonie; qu'elle en abrège la durée; que cependant cette influence est beaucoup moindre qu'on ne se l'imagine communément... " (p. 62)

His contemporaries were not all persuaded. According to one, arithmetic should not have been allowed to constrain the imagination:

> "En invoquant l'inflexibilité de l'arithmétique pour se soustraire aux empiétemens de l'imagination, on commet contre le bon sens la plus grave erreur... " (p. 79)

Pierre Louis was comparing average outcomes for patients bled early or late in the course of the disease. The critic felt that the groups were different in important respects apart from treatment. Louis replied that individual differences made it impossible to learn much from studying individual cases and necessitated the use of averages; see also Gavarret (1840). This tension has never been fully resolved, and is with us even today.

A few statistical details may be of interest. Louis reports on 78 pneumonia patients. All were bled, at different stages of the disease, and 50 survived. Among the survivors, bleeding in the first two days cut the length of the illness in half. But, Louis noted, there were differences in régime. Those treated later had not followed doctors' orders:

> "[ils] avaient commis des erreurs de régime, pris des boissons fortes, du vin chaud sucré, un ou plusieurs jours de suite, en quantité plus ou moins considérable; quelquefois même de l'eau-de-vie. " (p. 13)

From a modern perspective, there is a selection effect in Louis' analysis: those treated later in the course of an illness are likely for that reason alone to have had longer illnesses. It therefore seems better to consider outcomes for all 78 patients, including those who died, and bleeding in the first two days doubles the risk of death. Louis saw this, but dismissed it as frightening and absurd on its face:

> "Résultat effrayant, absurde en apparence. " (p. 17)

He explains that those who were bled later were older. He was also careful to point out the limitations created by a small sample.

Among other things, Louis identified two major methodological issues: (i) sampling error and (ii) confounding. These problems must be addressed in any epidemiologic study. Confounding is the more serious issue. In brief, a comparison is made between a treatment group and a control group, in order to determine the effect of treatment. If the groups differ with respect to another factor—the "confounding variable"—which influences the outcome, the estimated treatment effect will also include the effect of the confounder, leading to a potentially serious bias. If the treatment and control groups are chosen at random, bias is minimized. Of course, in epidemiologic studies, there are many other sources of bias besides confounding. One example is "recall bias," where a respondent's answers to questions about exposure are influenced by presence or absence of disease. Another example is "selection bias," due for instance to systematic differences between subjects chosen for a study and subjects excluded from the study. Even random measurement error can create bias in estimated effects: random errors in measuring the size of a causal factor tend to create a bias toward 0, while errors in measuring a confounder create a bias in the opposite direction.

Pierre Louis' book was published in the same year as Quetelet's *Sur l'homme et le développement de ses facultés, ou Essai de physique sociale.* Quetelet, like Louis, has had—and continues to have—an important influence over the development of our subject (Sections 4 and 5).

3. Snow on Cholera

In 1855, some twenty years before Koch and Pasteur laid the foundations of modern microbiology, Snow discovered that cholera is a waterborne infectious disease. At the time, the germ theory of disease was only one of many conceptions. Imbalance in the humors of the body was an older explanation for disease. Miasma, or bad air, was often said to be the cause of epidemics. Poison in the ground was perhaps a slightly later idea.

Snow was a physician in London. By observing the course of the disease, he concluded that cholera was caused by a living organism, which entered the body with water or food, multiplied in the body, and made the body expel water containing copies of the organism. The dejecta then contaminated food or reentered the water supply, and the organism proceeded to infect other victims. The lag between infection and disease (a matter of hours or days) was explained as the time needed for the infectious agent to multiply in the body of the victim. This multiplication is characteristic of life: inanimate poisons do not reproduce themselves.

Snow developed a series of arguments in support of the germ theory. For instance, cholera spread along the tracks of human commerce. Furthermore, when a ship entered a port where cholera was prevalent, sailors contracted the disease only when they came into contact with residents of the port. These facts were easily explained if cholera was an infectious disease, but were harder to explain by the miasma theory.

There was a cholera epidemic in London in 1848. Snow identified the first or "index" case in this epidemic:

> "a seaman named John Harnold, who had newly arrived by the *Elbe* steamer from Hamburgh, where the disease was prevailing. " (p. 3)

He also identified the second case: a man named Blenkinsopp who took Harnold's room after the latter died, and presumably became infected by contact with the bedding. Next, Snow was able to find adjacent apartment buildings, one being heavily affected by cholera and one not. In each case, the affected building had a contaminated water supply; the other had relatively pure water.

Again, these facts are easy to understand if cholera is an infectious disease, but hard to explain on the miasma theory.

There was an outbreak of the disease in August and September of 1854. Snow made what is now called a "spot map," showing the locations of the victims. These clustered near the Broad Street pump. (Broad Street is in Soho, London; at the time, public pumps were used as a source of water.) However, there were a number of institutions in the area with few or no fatalities. One was a brewery. The workers seemed to have preferred ale to water: but if any wanted water, there was a private pump on the premises. Another institution almost free of cholera was a poorhouse, which too had its own private pump. People in other areas of London contracted the disease; but in most cases, Snow was able to show they drank water from the Broad Street pump. For instance, one lady in Hampstead so much liked its taste that she had bottled water from the Broad Street pump delivered to her house by carter.

So far, we have persuasive anecdotal evidence that cholera is an infectious disease, spread by contact or through the water supply. Snow also used statistical ideas. There were a number of water companies in the London of his time. Some took their water from heavily contaminated stretches of the Thames river; for others, the intake was relatively uncontaminated. Snow made what are now called "ecological" studies, correlating death rates from cholera in various areas of London with the quality of the water. Generally speaking, areas with contaminated water had higher death rates. One exception was the Chelsea water company. This company started with contaminated water, but had quite modern methods of purification—settling ponds, exposure to sunlight, and sand filtration. Its service area had a low death rate from cholera.

In 1852, the Lambeth water company moved its intake pipe upstream to secure relatively pure water. The Southwark and Vauxhall company left its intake pipe where it was, in a heavily contaminated stretch of the Thames. Snow made an ecological analysis comparing the areas serviced by the two companies in the epidemics of 1853-54 and in earlier years. Let him now continue in his own words.

> "Although the facts shown in the above table [the ecological analysis] afford very strong evidence of the powerful influence which the drinking of water containing the sewage of a town exerts over the spread of cholera, when that disease is present, yet the question does not end here; for the intermixing of the water supply of the Southwark and Vauxhall Company with that of the Lambeth Company, over an extensive part of London, admitted of the subject being sifted in such a way as to yield the most incontrovertible proof on one side or the other. In the subdistricts enumerated in the above table as being supplied by both Companies, the mixing of the supply is of the most intimate kind. The pipes of each Company go down all the streets, and into nearly all the courts and alleys. A few houses are supplied by one Company and a few by the other, according to the decision of the owner or occupier at that time when the Water

5. FROM ASSOCIATION TO CAUSATION

Companies were in active competition. In many cases a single house has a supply different from that on either side. Each company supplies both rich and poor, both large houses and small; there is no difference either in the condition or occupation of the persons receiving the water of the different Companies. Now it must be evident that, if the diminution of cholera, in the districts partly supplied with improved water, depended on this supply, the houses receiving it would be the houses enjoying the whole benefit of the diminution of the malady, whilst the houses supplied with the [contaminated] water from Battersea Fields would suffer the same mortality as they would if the improved supply did not exist at all. As there is no difference whatever in the houses or the people receiving the supply of the two Water Companies, or in any of the physical conditions with which they are surrounded, it is obvious that no experiment could have been devised which would more thoroughly test the effect of water supply on the progress of cholera than this, which circumstances placed ready made before the observer.

"The experiment, too, was on the grandest scale. No fewer than three hundred thousand people of both sexes, of every age and occupation, and of every rank and station, from gentlefolks down to the very poor, were divided into groups without their choice, and in most cases, without their knowledge; one group being supplied with water containing the sewage of London, and amongst it, whatever might have come from the cholera patients; the other group having water quite free from such impurity.

"To turn this grand experiment to account, all that was required was to learn the supply of water to each individual house where a fatal attack of cholera might occur. " (pp. 74–75)

Snow's data are shown in Table 5.1. The denominator data—the number of houses served by each water company—were available from parliamentary records. For the numerator data, however, a house-to-house canvass was needed to determine the source of the water supply at the address of each cholera fatality. (The "bills of mortality" showed the address, but not the water source.) The death rate from the Southwark and Vauxhall water is about nine times the death rate for the Lambeth water. This is compelling evidence.

Snow argued that the data could be analyzed as if they had resulted from an experiment of nature: there was no difference between the customers of the two water companies, except for the water. His sample was not only large but representative; therefore, it was possible to generalize to a larger population. Finally, Snow was careful to avoid the "ecological fallacy": relationships that hold for groups may not hold for individuals (Robinson, 1950). It is the design of the study and the magnitude of the effect that compel conviction, not the elaboration of technique.

Table 5.1. Death rate from cholera by source of water. Rate per 10,000 houses. London, epidemic of 1853–54. Snow's Table IX.

	No. of Houses	Cholera Deaths	Rate per 10,000
Southwark & Vauxhall	40,046	1,263	315
Lambeth	26,107	98	37
Rest of London	256,423	1,422	59

More evidence was to come from other countries. In New York, the epidemics of 1832 and 1849 were handled according to the theories of the time. The population was exhorted to temperance and calm, since anger could increase the humor "choler" (bile), and imbalances in the humors of the body lead to disease. Pure water was brought in to wash the streets and reduce miasmas. In 1866, however, the epidemic was handled by a different method—rigorous isolation of cholera cases, with disinfection of their dejecta by lime or fire. The fatality rate was much reduced.

At the end of the 19th century, there was a burst of activity in microbiology. In 1878, Pasteur published *La théorie des germes et ses applications à la médecine et à la chirurgie*. Around that time, Pasteur and Koch isolated the anthrax bacillus and developed techniques for vaccination. The tuberculosis bacillus was next. In 1883, there was a cholera epidemic in Egypt, and Koch isolated the vibrio; he was perhaps anticipated by Filipo Pacini. There was an epidemic in Hamburg in 1892. The city fathers turned to Max von Pettenkofer, a leading figure in the German hygiene movement of the time. He did not believe Snow's theory, holding instead that cholera was caused by poison in the ground. Hamburg was a center of the slaughterhouse industry, and von Pettenkofer had the carcasses of dead animals dug up and hauled away, in order to reduce pollution of the ground. The epidemic continued its ravages, which ended only when the city lost faith in von Pettenkofer and turned in desperation to Koch.

The approach developed by Louis and Snow found many applications; I will mention only two examples. Semmelweiss (1867) discovered the cause of puerperal fever. Around 1914, Goldberger showed that pellagra was the result of a diet deficiency. Terris (1964) reprints many of Goldberger's articles; also see Carpenter (1981). A useful reference on Pasteur is Dubos (1988). References on the history of cholera include Rosenberg (1962), Howard-Jones (1975), Evans (1987), Winkelstein (1995), Paneth et al. (1998). Today, the molecular biology of the cholera vibrio is reasonably well understood; see, for instance, Finlay, Heffron, and Fialkow (1989) or Miller, Mekalanos, and Fialkow (1989). For a synopsis, see Alberts et al. (1994, pp. 484, 738); there are recent surveys by Colwell (1996) and Raufman (1998). Problems with ecological inference are discussed in Freedman (2001).

4. Regression Models in Social Science

Legendre (1805) and Gauss (1809) developed the regression method (least absolute residuals or least squares) to fit data on the orbits of astronomical objects. In this context, the relevant variables are known and so are the functional forms of the equations connecting them. Measurement can be done to high precision, and much is known about the nature of the errors- in the measurements and the equations. Furthermore, there is ample opportunity for comparing predictions to reality.

By the turn of the century, investigators were using regression on social science data where these conditions did not hold, even to a rough approximation. One of the earliest such papers is Yule (1899), "An investigation into the causes of changes in pauperism in England, chiefly during the last two intercensal decades." At the time, paupers were supported either inside "poor-houses" or outside, depending on the policy of local authorities. Did the relief policy affect the number of paupers? To study this question, Yule offered a regression equation,

$$\Delta \text{Paup} = a + b \times \Delta \text{Out} + c \times \Delta \text{Old} + d \times \Delta \text{Pop} + \text{error}.$$

In this equation,

> Δ is percentage change over time,
> "Out" is the out-relief ratio N/D,
> N = number on welfare outside the poor-house,
> D = number inside,
> "Old" is the percentage of the population over 65,
> "Pop" is the population.

Data are from the English Censuses of 1871, 1881, 1891. There are two Δ's, one for 1871–81 and one for 1881–91.

Relief policy was determined separately by the local authorities in each "union," a small geographical area like a parish. At the time, there were about 600 unions, and Yule divided them into four kinds: rural, mixed, urban, metropolitan. There are 2 × 4 = 8 equations, one for each combination of time period and type of union. Yule assumed that the coefficients were constant for each equation, which he fitted to the data by least squares. That is, he estimated the coefficients a, b, c, and d as the values that minimized the sum of squared errors,

$$\sum \left(\Delta \text{Paup} - a - b \times \Delta \text{Out} - c \times \Delta \text{Out} - d \times \Delta \text{Pop} \right)^2.$$

The sum is taken over all unions of a given type at a given time-period.

For example, consider the metropolitan unions. Fitting the equation to the data for 1871–81 gave

$$\Delta \text{Paup} = 13.19 + 0.755 \, \Delta \text{Out} - 0.022 \, \Delta \text{Old} - 0.322 \, \Delta \text{Pop} + \text{residual}.$$

For 1881–91, Yule's equation was

$$\Delta \text{Paup} = 1.36 + 0.324 \, \Delta \text{Out} + 1.37 \, \Delta \text{Old} - 0.369 \, \Delta \text{Pop} + \text{residual}.$$

The framework combines the ideas of Quetelet with the mathematics of Gauss. Yule is studying the "social physics" of poverty. Nature has run an experiment, assigning different treatments to different areas. Yule is analyzing the results, using regression to isolate the effects of out-relief. His principal conclusion is that welfare outside the poor-house creates paupers—the estimated coefficient on the out-relief ratio is positive.

At this remove, the flaws in the argument are clear. Confounding is a salient problem. For instance, Pigou (a famous economist of the era) thought that unions with more efficient administrations were the ones building poorhouses and reducing poverty. Efficiency of administration is then a confounder, influencing both the presumed cause and its effect. Economics may be another confounder. At times, Yule seems to be using the rate of population change as a proxy for economic growth, although this is not entirely convincing. Generally, however, he pays little attention to economic activity. The explanation:

> "A good deal of time and labor was spent in making trial of this idea, but the results proved unsatisfactory, and finally the measure was abandoned altogether. " (p. 253)

The form of his equation is somewhat arbitrary, and the coefficients are not consistent over time and space. This is not necessarily fatal. However, if the coefficients do not exist separately from the data, how can they predict the results of interventions? There are also problems of interpretation. At best, Yule has established association. Conditional on the covariates, there is a positive association between ΔPaup and ΔOut. Is this association causal? If so, which way do the causal arrows point? These questions are not answered by the data analysis; rather, the answers are assumed a priori. Yule is quite concerned to parcel out changes in pauperism: so much is due to changes in the out-relief ratio, so much to changes in other variables, and so much to random effects. However, there is one deft footnote (number 25) that withdraws all causal claims:

> "Strictly speaking, for 'due to' read 'associated with.'"

5. FROM ASSOCIATION TO CAUSATION

Yule's approach is strikingly modern, except there is no causal diagram and no stars indicating statistical significance. Figure 5.1 brings him up to date. An arrow from X to Y indicates that X is included in the regression equation that explains Y. "Statistical significance" is indicated by an asterisk, and three steriks signal a high degree of significance. The idea is that a statistically significant coefficient differs from 0, so that X has a causal influence on Y. By contrast, an insignificant coefficient is zero: then X does not exert a causal influence on Y.

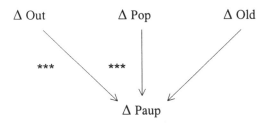

Figure 5.1. Yule's Model. Metropolitan unions, 1871–81.

The reasoning is seldom made explicit, and difficulties are frequently overlooked. Stringent assumptions are needed to determine significance from the data. Even if significance can be determined and the null hypothesis rejected or accepted, there is a much deeper problem. To make causal inferences, it must in essence be assumed that equations are invariant under proposed interventions. Verifying such assumptions—without making the interventions—is quite problematic. On the other hand, if the coefficients and error terms change when the right hand side variables are manipulated rather than being passively observed, then the equation has only a limited utility for predicting the results of interventions. These difficulties are well known in principle, but are seldom dealt with by investigators doing applied work in the social and life sciences. Despite the problems, and the disclaimer in the footnote, Yule's regression approach has become widely used in the social sciences and epidemiology.

Some formal models for causation are available, starting with Neyman (1923). See Hodges and Lehmann (1964, sec. 9.4), Rubin (1974), or Holland (1988). More recent developments will be found in Pearl (1995; 2000) or Angrist, Imbens and Rubin (1996). For critical discussion from various perspectives, see Goldthorpe (1998; 2001), Humphreys and Freedman (1996, 1999), Abbott (1997), McKim and Turner (1997), Manski (1995), Lieberson (1985), Lucas (1976), Liu (1960), or Freedman (1987; 1991; 1995). Ní Bhrolcháin (2001) presents some fascinating case studies. The role of invariance is considered in Heckman (2000) and Freedman (2002). The history is reviewed by Stigler (1986) and Desrosières (1993).

5. Regression Models in Epidemiology

Regression models (and variations like the Cox model) are widely used in epidemiology. The models seem to give answers, and create at least the appearance of methodological rigor. This section discusses one example, which is fairly typical of such applications and provides an interesting contrast to Snow on cholera. Snow used primitive statistical techniques, but his study designs were extraordinarily well thought out, and he made a huge effort to collect the relevant data. By contrast, many empirical papers published today, even in the leading journals, lack a sharply-focused research question; or the study design connects the hypotheses to the data collection only in a very loose way. Investigators often try to use statistical models not only to control for confounding, but also to correct basic deficiencies in the design or the data. Our example will illustrate some of these points.

Kanarek et al. (1980) asked whether asbestos in the drinking water causes cancer. They studied 722 census tracts in the San Francisco Bay Area. (A census tract is a small geographical region, with several thousand inhabitants.) The investigators measured asbestos concentration in the water for each tract. Perhaps surprisingly, there is enormous variation; less surprisingly, higher concentrations are found in poorer tracts. Kanarek et al. compared the "observed" number of cancers by site with the expected number, by sex, race, and tract. The "expected" number is obtained by applying age-specific national rates to the population of the tract, age-group by age-group; males and females are done separately, and only Whites are considered. (There are about 100 sites for which age-specific national data are available; comparison of observed to expected numbers is an example of "indirect standardization.")

Regression is used to adjust for income, education, marital status, and occupational exposure. The equation is not specified in great detail, but is of the form

$$\log \frac{\text{Obs.}}{\text{Exp.}} = A_0 + A_1 \text{ asbestos concentration} + A_2 \text{ income} + A_3 \text{ education} + A_4 \text{ married} + A_5 \text{ asbestos workers} + \text{error}.$$

Here, "income" is the median figure for persons in the tract, and "education" is the median number of years of schooling; data are available from the census. These variables adjust to some extent for socioeconomic differences between tracts: usually, rates of disease go down as income and education go up. The next variable in the equation is the fraction of persons in the tract who are married; such persons are typically less subject to disease than the unmarried. Finally, there is the number of "asbestos workers" in the tract; these persons may have unusually high rates of cancer, due to exposure on the job. Thus, the variables on the right hand side of the equation are potential confounders, and the equation tries to adjust for their effects. The estimate of A_1 for lung cancer in males is "highly statistically significant," with $P < .001$. A highly significant coefficient like this might be taken as evidence of causation, but there are serious difficulties.

Confounding. No adjustment is made for smoking habit, which was not measured in this study. Smoking is strongly but imperfectly associated with socioeconomic status, and hence with asbestos concentration in the water. Furthermore, smoking has a substantial effect on cancer rates. Thus, smoking is a confounder. The equation does not correct for the effects of smoking, and the P-value does not take this confounding into account.

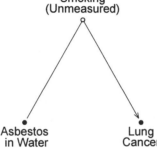

Figure 5.2. Smoking as an unmeasured confounder. The non-causal association between asbestos in the water and lung cancer is explained by the associations with smoking.

Figure 5.2 illustrates an alternative explanation for the data. (i) Smoking (an unmeasured confounder) is associated with the concentration of asbestos fibers in the water. The association is signaled by the straight line joining the two variables. (ii) Smoking has a strong, direct effect on lung cancer, indicated by the arrow in the figure. Associations (i) and (ii) explain the association between asbestos fibers in the water and lung cancer rates; this observed association is not causal. To recapitulate, a confounder is associated with the putative cause and with its effect; the confounder may explain part or all of an observed association. In epidemiology, unmeasured or poorly measured confounders are the rule rather than the exception. (Technically, the relationships in Figure 5.2 must persist even after conditioning on the measured covariates.)

Model specification. The choice of variables and functional form is somewhat arbitrary, although not completely unreasonable. The authors say that their equation is suggested by mathematical models for cancer, but the connection is rather loose; nor have the cancer models themselves been validated (Freedman & Navidi, 1989; 1990).

Statistical assumptions. To compute the *P*-value, it is tacitly assumed that errors are statistically independent from tract to tract, and identically distributed. This assumption may be convenient, but it lacks an empirical basis.

The search for significance. Even if we set the fundamental difficulties aside, the authors have made several hundred tests on the equations they report, without counting any preliminary data analysis that may have been done. The *P*-values are not adjusted for the effects of the search, which may be substantial (Dijkstra, 1988; Freedman, 1983).

Weak effects. The effect being studied is weak: a 100-fold increase in asbestos fiber concentration is associated with perhaps a 5% increase in lung cancer rates. What is unusual about this example is only the strength of the unmeasured confounder, and the weakness of the effect under investigation.

Epidemiology is best suited to the investigation of strong effects, which are hard to explain away by confounding (Cornfield et al., 1959, p. 199). As attention shifts to the weaker and less consistent effects that may be associated with low doses, difficulties will increase. Long delays between the beginning of exposure and the onset of disease are a further complication. Toxicology may be of some value but presents difficulties of its own (Freedman, Gold, & Lin, 1996; Freedman & Zeisel, 1988). The limitations of epidemiology are discussed by Taubes (1995). For detailed case studies, see Vandenbroucke and Pardoel (1989), Taubes (1998), or Freedman and Petitti (2001). Other examples will be given in Section 7.

6. Some General Considerations

Model specification. A model is specified by choosing (i) the explanatory variables to put on the right-hand side, (ii) the functional form of the equation, and (iii) the assumptions about error terms. Explanatory variables are also called "covariates," or "independent variables"; the latter term does not connote statistical independence. The functional form may be linear, or log linear, or something even more exotic. Errors may be assumed independent or autoregressive; or some other low-order covariance matrix may be assumed, with a few parameters to estimate from the data.

Epidemiologists often have binary response variables: for instance, disease is coded as "1" and health as "0". A "logit" specification is common in such circumstances. Conditional on the covariates, subjects are assumed to be independent. If Y_i is the response for subject i while X_i is a $1 \times p$ vector of covariates, the logit specification is

$$\log \frac{\text{Prob}\{Y_i = 1\}}{\text{Prob}\{Y_i = 0\}} = X_i \beta.$$

Here, β is a $p \times 1$ vector of parameters, which would be estimated from the data by maximum likelihood. For useful details on various models and estimation procedures, see Breslow and Day (1980; 1987).

5. FROM ASSOCIATION TO CAUSATION

Models are chosen on the basis of familiarity and convenience; there will be some effort made to avoid gross conflict with the data. Choices are generally somewhat arbitrary, although they may not be unreasonable. There will often be some preliminary data analysis: for instance, variables with insignificant coefficients are discarded, and the model refitted. Details can make a large difference in conclusions. In particular, P-values are often strongly dependent on the final specification, and the preliminary screening may make these P-values difficult to interpret - as will be discussed.

It is sometimes argued that biases (like recall bias or selection bias) can be modeled and then corrections can be made. That might be so if the auxiliary models could themselves be validated. On the other hand, if the auxiliary models are of doubtful validity, the "corrections" they suggest may make matters worse rather than better. For more discussion, see Scharfstein, Rotnitzky and Robins (1999) or Copas and Li (1997). In the original physical-science applications, the specifications were dictated by prior theory and empirical fact (Section 4). In the social sciences and epidemiology, the specifications are much more arbitrary. That is a critical distinction, as discussed in Freedman (2002).

A review of P-values. It may be enough to consider one typical example. Suppose X is a random variable, distributed as N(μ, 1), so

$$\text{Prob}\{X - \mu < x\} = \Phi(x) = \frac{1}{\sqrt{2\pi}} \int_{-\infty}^{x} \exp(-u^2/2)\, du.$$

The "null hypothesis" is that $\mu = 0$; the "alternative" is that $\mu = 0$. The "test statistic" is $|X|$. Large values of the test statistic are evidence against the null hypothesis. For instance, a value of 2.5 for $|X|$ would be quite unusual- if the null hypothesis is correct. Such large values are therefore evidence against the null.

If x is the "observed value" of X, that is, the value realized in the data, then the P-value of the test is $\Phi(-|x|) + 1 - \Phi(|x|)$. In other words, P is the chance of getting a test statistic as large as or larger than the observed one; this chance is computed on the basis of the null hypothesis. (Sometimes, P is called the "observed significance level.") If the null hypothesis is correct, then P has a uniform distribution. Otherwise, P is more concentrated near 0. Thus, small values of P argue against the null hypothesis. If $P < .05$, the result is "statistically significant"; if $P < .01$, the result is "highly significant." These distinctions are somewhat arbitrary, but have a powerful influence on the way statistical studies are received. In this example, X is an unbiased estimate of μ. If X were biased, the bias would have to be estimated from some other data, and removed from X before proceeding with the test. P is about sampling error, not bias.

The search for significance. The effect of multiple comparisons can be seen in our example. A value of 2.5 for $|X|$ is unusual. However, if 1000 independent copies of X are examined, values of 2.5 or larger are to be expected. If only the large values are noticed and the search effort is ignored when computing P, severe distortion can result. Disease clusters attributed to environmental pollution may present such analytical problems. There are many groups of people, many sources of pollution, many possible routes of exposure, and many possible health effects. Great care is needed to distinguish real effects from the effects of chance. The search effort may not be apparent, because a cluster—especially of a rare disease—can be quite salient.

The difficulty is not widely appreciated, so another example may be useful. A coin that lands heads 10 times in a row is unusual. On the other hand, if a coin is tossed 1000 times and there is at least one run of 10 heads, that is only to be expected. The latter model may be more relevant for a disease cluster, given the number of possibilities open to examination.

If adjustment for confounding is done by regression and various specifications are tried, chance capitalization again comes into play. For some empirical evidence, see Ottenbacher (1998) or Dickersin (1997). Many epidemiologists deny that problems are created by the search for significance. Some commentators are more concerned with loss of power than distortions of the P-value, because they are convinced a priori that the null hypothesis is untenable. Of course, it is then unclear why statistical testing and P are relevant. See, for instance, Rothman (1990) or Perneger (1998). On the other hand, Rothman's preference for estimation over testing in the epidemiologic context often seems justified, especially when there is an effect to be estimated. For more discussion, see Cox (1977, Section 5), or Freedman, Pisani, and Purves (1997, chapter 29); also see section 9.

Intermediate variables. If X and Y cause Z, but X also causes Y, the variable Y would often be treated as an "intermediate variable" along the pathway from X to Z, rather than a confounder. If the object is to estimate the total effect of X on Z, then controlling for Y is usually not advised. If the idea to estimate the direct effect of X on Z, then controlling for Y may be advised, but the matter can under some circumstances be quite delicate. See Greenland, Pearl, and Robins (1999).

7. Other Examples in Epidemiology

This section provides more examples in epidemiology. Generally, the studies mentioned are unpersuasive, for one or more of the following reasons.
- Effects are weak and inconsistent.
- Endpoints are poorly defined.
- There is an extensive search for statistical significance.
- Important confounders are ignored.

When effects are weak or inconsistent, chance capitalization and confounding are persistent issues; poorly-defined endpoints lend themselves to a search for significance. These problems are particularly acute when studying clusters. However, the section ends on a somewhat positive note. After numerous false starts, epidemiology and molecular biology have identified the probable etiologic agent in cervical cancer.

Leukemias and sarcomas associated with exposure to electromagnetic fields. Many studies find a weak correlation between exposure to electromagnetic fields and a carcinogenic response. However, different studies find different responses in terms of tissue affected. Nor is there much consistency in measurement of dose, which would in any event be quite difficult. Some investigators try to measure dose directly, some use distance from power lines, some use "wire codes," which are summary measures of distance from transmission lines of different types. Some consider exposure to household appliances like electric blankets or microwave ovens, while some do not. The National Research Council (1997) reviewed the studies and concluded there was little evidence for a causal effect. However, those who believe in the effect continue to press their case.

Air pollution. Some investigators find an effect of air pollution on mortality rates: see Pope, Schwartz, and Ransom (1992). Styer et al. (1995) use similar data and a similar modeling strategy, but find weak or inconsistent effects; also see Gamble (1998). Estimates of risk may be determined largely by unverifiable modeling assumptions rather than data, although conventional opinion now finds small particles to be hazardous.

Sellafield. There was a leukemia cluster associated with the British nuclear facility at Sellafield. Fathers working in the facility were exposed to radiation, which was said to have damaged the sperm and caused cancer in the child after conception—the "paternal preconception irradiation" hypothesis. Two of the Sellafield leukemia victims filed suit. There was a trial with discovery and cross examination of expert witnesses, which gives a special perspective on the epidemiology. As it turned out, the leukemia cluster had been discovered by reporters. The nature and intensity of the search is unknown; *P*-values were not adjusted for multiple comparisons. The effects of news stories on subsequent responses to medical interviews must also be a concern. The epidemiologists who investigated the cluster used a case-control design, but changed the definitions of cases and controls part way through the study. For such reasons among others, causation does not seem to have been demonstrated. The judge found that:

> "the scales tilt decisively in favor of the defendants and the plaintiffs therefore have failed to satisfy me on the balance of probabilities that [paternal preconception irradiation] was a material contributory cause of the [Sellafield] excess.... " (p. 209)

The cases are Reay and Hope v. British Nuclear Fuels, 1990 R No. 860, 1989 H No. 3689. Sellafield is also referred to as Seascale or Windscale in the opinion, written by the Honorable Mr. Justice French of the Queen's Bench. The epidemiology is reported by Gardner et al. (1990) and Gardner (1992); also see Doll, Evans, and Darby (1994). Case-control studies will be discussed again in the next section. Chance capitalization is not a fully satisfactory explanation for the Sellafield excess. Some epidemiologists think that leukemia clusters around nuclear plants may be a real effect, caused by exposure of previously isolated populations to viruses carried by immigrants from major population centers; this hypothesis was first put forward in another context by Kinlen and John (1994).

Cervical cancer. This cancer has been studied for many years. Some investigators have identified the cause as tissue irritation; others point to syphilis, or herpes, or chlamydia; still others have found circumcision of the husband to be protective. See Gagnon (1950), Røjel (1953), Aurelian et al. (1973), Hakama et al. (1993), or Wynder et al. (1954). Today, it is believed that cervical cancer is in large part a sexually transmitted disease, the agent being certain types of human papillomavirus, or HPV. There is suggestive evidence for this proposition from epidemiology and from clinical practice, as well as quite strong evidence from molecular biology. If so, the earlier investigators were misled by confounding. For example, the women with herpes were presumably more active sexually, and more likely to be exposed to HPV. The two exposures are associated, but it is HPV that is causal. For reviews, see Storey et al. (1998) or Cannistra and Niloff (1996). The history is discussed by Evans (1993, pp. 101–105); some of the papers are reprinted by Buck et al. (1989).

8. Health Effects of Smoking

In the 1920s, physicians noticed a rapid increase of death rates from lung cancer. For many years, it was debated whether the increase was real or an artifact of improvement in diagnostics. (The lungs are inaccessible, and diagnosis is not easy.) By the 1940s, there was some agreement on the reality of the increase, and the focus of the discussion shifted. What was the cause of the epidemic? Smoking was one theory. However, other experts thought that emissions from gas works were the cause. Still others believed that fumes from the tarring of roads were responsible.

Two early papers on smoking and lung cancer were Lombard and Doering (1928) and Müller (1939). Later papers attracted more attention, especially Wynder and Graham (1950) in the U.S. and Doll and Hill (1950; 1952) in the U.K. I will focus on the last, which reports on a "hospital-based case-control study." Cases were patients admitted to certain hospitals with a diagnosis of lung cancer; the controls were patients admitted for other reasons. Patients were interviewed about their exposure to cigarettes, emissions from gas works, fumes from tarring of the roads, and various other possible etiologic agents. Interviewing was done "blind," by persons unaware of the purpose of the study. The cases and controls turned out to have rather similar exposures to suspect agents- except for smoking. Data on that exposure are shown in Table 5.2.

5. FROM ASSOCIATION TO CAUSATION

Table 5.2. Hospital-based case-control study. Smoking status for cases and controls. From Doll and Hill (1952).

	Cases	Controls
Smoker	1350	1296
Nonsmoker	7	61

There were 1357 cases in the study, of whom 1350 were smokers; there were 1357 controls, of whom 1296 were smokers. In both groups, non-smokers are rare; but they are much rarer among the controls. To summarize such data, epidemiologists use the "odds ratio,"

$$\frac{1350/7}{1296/61} \approx 9.$$

Roughly speaking, lung cancer is 9 times more common among smokers than among non-smokers. (Doll and Hill matched their cases and controls, a subtlety that will be ignored here.) Interestingly enough, there are some cases where the diagnosis of lung cancer turned out to be wrong; these cases smoked at the same rate as the controls—an unexpected test confirming the smoking hypothesis.

The odds ratio is a useful descriptive statistic on its own. However, there is a conventional way of doing statistical inference in this setting, which leads to confidence intervals and P-values. The basic assumption is that the cases are a random sample from the population of lung cancer cases, while the controls are a random sample (with a different sampling fraction) from the part of the population that is free of the disease. The odds ratio in the data would then estimate the odds ratio in the population.

Table 5.3. A 2×2 table for the population, classified according to presence or absence of lung cancer and smoking habit: a is the number of smokers with lung cancer, b is the number of smokers free of the disease, and so forth.

	Lung cancer	No lung cancer
Smoker	a	b
Nonsmoker	c	d

More explicitly, the population can be classified in a 2×2 table, as in Table 5.3, where a is the number who smoke and have lung cancer; b is the number who smoke but do not have lung cancer; similarly for c and d. Suppose the lung cancer patients in hospital are sampled at the rate ϕ from the corresponding part of the population, while the controls are sampled at the rate ψ from the remainder of the population. With a large number of patients, the odds ratio in the study is essentially the same as the odds ratio in the population (Cornfield, 1951), because

$$\frac{(\phi a)/(\phi c)}{(\psi b)/(\psi d)} = \frac{a/c}{b/d}.$$

Since lung cancer is a rare disease even among smokers, $a/b \approx a/(a+b)$ approximates the rate of disease among smokers, while $c/d \approx c/(c+d)$ approximates the rate among non-smokers, and the odds ratio nearly coincides with the rate ratio. Moreover, standard errors and the like can be computed on the basis of the sampling model. For details, see Breslow and Day (1980).

The realism of the model, of course, is open to serious doubt: patients are not hospitalized at random. This limits the usefulness of confidence intervals and P-values. Scientifically, the strength of the case against smoking rests not so much on the P-values, but more on the size of the effect, its coherence, and on extensive replication both with the original research design and with many other designs. Replication guards against chance capitalization and, at least to some extent, against confounding— if there is some variation in study design (Cornfield et al., 1959; Ehrenberg & Bound, 1993).

The epidemiologists took full advantage of replication. For instance, Doll and Hill (1954) began a "cohort study," where British doctors were followed over time and mortality rates were studied in relation to smoking habit. At this point, it became clear that the smokers were dying at much faster rates than the non-smokers, not only from lung cancer but from many other diseases, notably coronary heart disease. It also became clear that the odds ratio computed from Table 5.2 was biased downward, because patients in a hospital are more likely to be smokers than the general population.

Coherence of study results is also an important part of the case; (i) There is a dose-response relationship: persons who smoke more heavily have greater risks of disease than those who smoke less; (ii) The risk from smoking increases with the duration of exposure; (iii) Among those who quit smoking, excess risk decreases after exposure stops. These considerations are systematized to some degree by "Hill's postulates:" see Evans (1993, pp. 186ff). Of course, the data are not free of all difficulties. Notably, inhalation increases the risk of lung cancer only in some of the studies.

There was resistance to the idea that cigarettes could kill. The list of critics was formidable, including Berkson (1955) and Fisher (1959); for a summary of Fisher's arguments, see Cook (1980). The epidemiologists made an enormous effort to answer criticisms and to control for possible confounders that were suggested. To take only one example, Fisher advanced the "constitutional hypothesis" that there was a genetic predisposition to smoke and to have lung cancer: genotype is the confounder. If so, there is no point in giving up cigarettes, because the risk comes from the genes not the smoke. To refute Fisher, the epidemiologists studied monozygotic twins. The practical difficulties are considerable, because we need twin pairs where one smokes and the other does not; furthermore, at least one of the twins must have died from the disease of interest. Monozygotic twins are scarce, smoking-discordant twin pairs scarcer yet. And lung cancer is a very rare disease, even among heavy smokers.

Data from the Finnish twin study (Kaprio & Koskenvuo, 1989) are shown in Table 5.4. There were 22 smoking-discordant monozygotic twin pairs where at least one twin died. In 17 out of 22 cases, the smoker died first. Likewise, there were 9 cases where at least one twin in the pair died of coronary heart disease. In each case, the smoker won the race to death. For all-cause mortality or coronary heart disease, the constitutional hypothesis no longer seems viable. For lung cancer, the numbers are tiny. Of course, other studies could be brought into play (Carmelli & Page, 1996). The epidemiologists refuted Fisher by designing appropriate studies and collecting the relevant data, not by a priori arguments and modeling. For other views, see Bross (1960) or Stolley (1991).

Table 5.4. The Finnish twin study. First death by smoking status among smoking-discordant twin pairs. From Kaprio and Koskenvuo (1989).

	Smokers	Non-smokers
All causes	17	5
Coronary heart disease	9	0
Lung cancer	2	0

Figure 5.3 shows current data from the U.S., with age-standardized death rates for the six most common cancers among males. Cancer is a disease of old age and the population has been getting steadily older, so standardization is essential. In brief, 1970 was chosen as a reference population. To get the standardized rates, death rates for each kind of cancer and each age group in each year are applied to the reference population. Mathematically, the standardized death rate from cancer of type j in year t is

$$\frac{\sum_i n_i d_{ijt}}{\sum_i n_i},$$

where n_i is the number of men in age group i in the 1970 population, and d_{ijt} is the death rate from cancer of type j among men in age group i in the population corresponding to year t. That is "direct standardization."

As will be seen, over the period 1930–80, there is a spectacular increase in lung cancer rates. This seems to have followed by about 20 or 25 years the increase in cigarette smoking. The death rate from lung cancer starts turning down in the late 1980s, because cigarette smoking began to decrease in the late 1960s. Women started smoking later than men, and continued longer: their graph (not shown) is lower, and still rising. The data on U.S. cigarette consumption are perhaps not quite as solid as one might like; for English data, which tell a very similar story, see Doll (1987) and Wald and Nicolaides-Bouman (1991). The initial segment of the lung cancer curve in Figure 5.3 was one of the first clues in the epidemiology of smoking. The downturn in the 1980s is one of the final arguments on the smoking hypothesis.

The strength of the case rests on the size and coherence of the effects, the design of the underlying epidemiologic studies, and on replication in many contexts. Great care was taken to exclude alternative explanations for the findings. Even so, the argument requires a complex interplay among many lines of evidence. Regression models are peripheral to the enterprise. Cornfield et al. (1959) provides an interesting review of the evidence in its early stages. A summary of more recent evidence will be found in IARC (1986). Gail (1996) discusses the history.

Figure 5.3. Age-standardized cancer death rates for males, 1930–1994. Per 100,000. U.S. vital statistics.

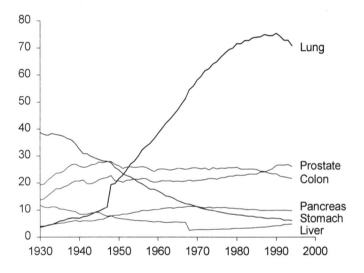

Note: Reprinted by the permission of the American Cancer Society, Inc. Figure is redrawn from American Cancer Society (1997), using data kindly provided by the ACS. According to the ACS, "Due to changes in ICD coding, numerator information has changed over time. Rates for cancers of the liver, lung, and colon and rectum are affected by these coding changes. Denominator information for the years 1930–1959 and 1991–1993 is based on intercensal population estimates, while denominator information for the years 1960–1989 is based on postcensal recalculation of estimates. Rate estimates for 1968–1989 are most likely of better quality."

9. Other Views

According to my near-namesake Friedman (1953, p. 15), "the relevant question to ask about the 'assumptions' of a theory is not whether they are descriptively 'realistic,' for they never are, but ... whether the theory works, which means whether it yields sufficiently accurate predictions". This argument is often used by proponents of modeling. However, the central question has been begged: how do we know whether the model is making good predictions? Fitting an equation to an existing data set is one activity; predicting the results of an intervention is quite another, and the crucial issue is getting from here to there. If regression models were generally successful in making causal inferences from associational data, that would be compelling evidence. In my experience, however, those who deploy Friedman's argument are seldom willing to engage in detailed discussions of the track record. Their reluctance is understandable.

Bross (1960, p. 394) writes, "a critic who objects to a bias in the design [of a study] or a failure to control some established factor is, in fact, raising a counterhypothesis ... [and] has the responsibility for showing that his counterhypothesis is tenable. In doing so, he operates under the same ground rules as [the] proponent." Also see Blau and Duncan (1967, p. 175). There is some merit to this point. Critics, like others, have an obligation to be reasonable. However, the argument is often used to shift the burden of proof from the proponent of a theory to the critic. That is perverse. Snow and his peers sought to carry the burden of proof, not to shift it. That is why their discoveries have stood the test of time.

Some observers concede that regression models can be misleading in attempts to identify causes; once causation has been established, however, they believe the models can be used to quantify the effects. Quantification is by no means straightforward. It is not only causation that must be established, but also the specification of the model, including the identification of the principal confounders, and the form of the equation connecting the relevant factors to the outcomes of interest (Section 6). The number of successes under this heading is not large.

Rothman and others have expressed a preference for confidence intervals over hypothesis testing. There have been objections, on the grounds that the two forms of inference are isomorphic. These objections miss the point. The isomorphism can tell us how to translate one set of mathematical theorems into another, but can scarcely dictate the form of an empirical research question. An investigator may be interested in a point estimate for some parameter, and may also want a measure of the uncertainty due to random error. For such an investigator, testing a sharp null hypothesis may be irrelevant. That would lead to confidence intervals, not *P*-values. Such an investigator, of course, would not care whether the confidence interval just misses—or just covers—some critical value, like 1.0 for an odds ratio.

To justify his position, Rothman makes two arguments: (i) fixed-level significance testing often creates artificial dichotomies; (ii) practitioners find it easier to misinterpret P-values than point estimates. See Rothman (1996), Lang, Rothman, and Cann (1998), or Rothman and Greenland (1998, pp. 183–194). Of course, objections to P-values can be taken to extremes: when Rothman was editor of *Epidemiology,* he banished P-values from the pages of that journal.

10. Summary and Conclusions

Statisticians generally prefer to make causal inferences from randomized controlled experiments, using the techniques developed by Fisher and Neyman. In many situations, experiments are impractical or unethical. Most of what we know about causation in such contexts is derived from observational studies. Sometimes, these are analyzed by regression models; sometimes, these are treated as natural experiments, perhaps after conditioning on covariates. Delicate judgments are required in order to assess the probable impact of confounders (measured and unmeasured), other sources of bias, and the adequacy of the statistical models used to make adjustments. There is much room for error in this enterprise, and much room for legitimate disagreement.

Snow's work on cholera, among other examples, shows that causal inferences can be drawn from non-experimental data. However, no mechanical rules can be laid down for making such inferences; since Hume's day, that is almost a truism. Indeed, causal inference seems to require an enormous investment of skill, intelligence, and hard work. Many convergent lines of evidence must be developed. Natural variation needs to be identified and exploited. Data must be collected. Confounders need to be considered. Alternative explanations have to be exhaustively tested. Above all, the right question needs to be framed.

Naturally, there is a desire to substitute intellectual capital for labor. That is why investigators often try to base causal inference on statistical models. With this approach, P-values play a crucial role. The technology is relatively easy to use, and promises to open a wide variety of questions to the research effort. However, the appearance of methodological rigor can be deceptive. Like confidence intervals, P-values generally deal with the problem of sampling error not the problem of bias. Even with sampling error, artifactual results are likely if there is any kind of search over possible specifications for a model, or different definitions of exposure and disease.

The modeling approach itself demands critical scrutiny. Mathematical equations are used to adjust for confounding and other sources of bias. These equations may appear formidably precise, but they typically derive from many somewhat arbitrary choices. Which variables to enter in the regression? What functional form to use? What assumptions to make about error terms? These choices are seldom dictated either by data or prior scientific knowledge. That is why judgment is so critical, the opportunity for error so large, and the number of successful applications so limited.

Acknowledgments

I would like to thank a number of people for useful comments, including Henri Causinus, Mike Finkelstein, Mark Hansen, Ben King, Erich Lehmann, Roger Purves, Ken Rothman, and Terry Speed. This chapter is based on a lecture I gave at the Académie des Sciences, Paris in 1998, on my Wald Lectures in Dallas later that year, and on lectures at the Athens University of Economics and Business in 2001. Sections 3 and 5 are adapted from Freedman (1991); section 4, from Freedman (1997). Previous versions of this chapter were published in *Statistical Science* (1999) and *Journal de la Société Française de Statistique* (2000).

11. References

Abbott, A. (1997). Of time and space: the contemporary relevance of the Chicago school. *Social Forces*, 75, 1149-1182.

Alberts, B., Bray, D., Lewis, J., Raff, M., Roberts, K. and Watson, J. D. (1994). *Molecular Biology of the Cell*, 3rd. ed., Garland Publishing, New York.

American Cancer Society (1997). *Cancer Facts & Figures-1997*. Atlanta, Georgia.

Angrist, J. D., Imbens, G. W. and Rubin, D. B. (1996). Identification of causal effects using instrumental variables. *J. Amer. Statist. Assoc.*, 91, 444-472.

Aurelian, L., Schumann, B., Marcus, R. L. and Davis, H. J. (1973). Antibody to HSV-2 induced tumor specific antigens in serums from patients with cervical carcinoma. *Science*, 181, 161-164.

Berkson, J. (1955). The statistical study of association between smoking and lung cancer. *Proc. Mayo Clinic*, 30, 319-348.

Blau, P. M. and Duncan, O. D. (1967). *The American Occupational Structure*. Wiley, New York. Chapter 5.

Breslow, N. and Day, N. E. (1980). *Statistical Methods in Cancer Research*, Vol. 1. International Agency for Research on Cancer, Lyon. Sci. Publ. No. 32. Distributed by Oxford University Press.

Breslow, N. and Day, N. E. (1987). *Statistical Methods in Cancer Research*, Vol. 2, International Agency for Research on Cancer, Lyon. Sci. Publ. No. 82. Distributed by Oxford University Press.

Bross, I. D. J. (1960). Statistical criticism. *Cancer*, 13, 394-400.

Buck, C., Llopis, A., Nájera, E. and Terris, M., eds. (1989). *The Challenge of Epidemiology: Issues and Selected Readings*, Sci. Publ. No. 505, World Health Organization, Geneva.

Cannistra, S. A. and Niloff, J. M. (1996). Cancer of the uterine cervix. *New Engl. J. Med.*, 334, 1030-1038.

Carmelli, D. and Page, W. F. (1996). Twenty-four year mortality in World War II US male veteran twins discordant for cigarette smoking. *Int. J. Epidemiol,*. 25, 554-559.

Carpenter, K. J. (1981). *Pellagra*, Academic Press. New York.

Colwell, R. R. (1996). Global climate and infectious disease: the cholera paradigm. *Science*, 274, 2025-2031.

Cook, D. (1980). Smoking and lung cancer. In S. E. Fienberg and D. V. Hinkley, (eds.) *R. A. Fisher, An Appreciation.* Lecture notes in statistics, (Vol. 1, pp. 182-191). Springer-Verlag, New York.

Copas, J. B. and Li, H. G. (1997). Inference for non-random samples. *J. Roy. Statist. Soc. Ser., B*, 59, 55-77.

Cornfield, J. (1951). A method for estimating comparative rates from clinical data. Applications to cancer of the lung, breast and cervix. *J. Nat. Cancer Inst.*, 11, 1269-1275.

Cornfield, J., Haenszel, W., Hammond, E. C., Lilienfeld, A. M., Shimkin, M. B. and Wynder, E. L. (1959). Smoking and lung cancer: recent evidence and a discussion of some questions. *J. Nat. Cancer Inst.*, 22, 173-203.

Cox, D. (1977). The role of significance tests. *Scand. J. Statist.*, 4, 49-70.
Desrosières, A. (1993). *La politique des grands nombres: histoire de la raison statistique.* Editions La Découverte, Paris. English translation by C. Naish (1998). *The Politics of Large Numbers: A History of Statistical Reasoning.* Harvard University Press. Cambridge.
Dickersin, K. (1997). How important is publication bias? A synthesis of available data. *AIDS Education and Prevention*, 9 Suppl. A, 15-21.
Dijkstra, T. K., ed. (1988). *On Model Uncertainty and its Statistical Implications.* Lecture Notes No. 307 in Economics and Mathematical Systems, Springer.
Doll, R. (1987). Major epidemics of the 20th century: from coronary thrombosis to AIDS. *J. Roy. Statist. Soc. Ser. A*, 150, 373-395.
Doll, R., Evans, H. J. and Darby, S. C. (1994). Paternal exposure not to blame. *Nature*, 367, 678-680.
Doll, R. and Hill, A. B. (1950). Smoking and carcinoma of the lung: preliminary report. *Br. Med. J.*, ii, 1271-1286.
Doll, R. and Hill, A. B. (1952). A study of the aetiology of carcinoma of the lung. *Br. Med. J.*, ii, 739-748.
Doll, R. and Hill, A. B. (1954). The mortality of doctors in relation to their smoking habit: A preliminary report. *Br. Med. J.*, i, 1451-1455.
Dubos, R. (1988). *Pasteur and Modern Science.* Springer.
Ehrenberg, A. S. C. and Bound, J. A. (1993). Predictability and prediction. *J. Roy. Statist. Soc. Ser. A*, 156 Part 2, 167-206 (with discussion).
Evans, A. S. (1993). *Causation and Disease: A Chronological Journey.* Plenum, New York.
Evans, R. J. (1987). *Death in Hamburg: Society and Politics in the Cholera Years.* Oxford University Press. Oxford.
Finlay, B. B., Heffron, F. and Fialkow, S. (1989). Epithelial cell surfaces induce *Salmonella* proteins required for bacterial adherence and invasion. *Science*, 243, 940-942.
Fisher, R. A. (1959). *Smoking: The Cancer Controversy.* Oliver and Boyd, Edinburgh.
Freedman, D. (1983). A note on screening regression equations. *Amer. Statistician*, 37, 152-155.
Freedman, D. (1987). As others see us: a case study in path analysis. *J. Educational Statistics*, 12, 101-223.
Freedman, D. (1991). Statistical models and shoe leather. In P. Marsden, (ed.), *Sociological Methodology.*.
Freedman, D. (1995). Some issues in the foundation of statistics. *Foundations of Science*, 1, 19-83.
Freedman, D. (1997). From association to causation via regression. *Adv. Appl. Math.*, 18, 59-110.
Freedman, D. (2001). Ecological inference and the ecological fallacy. Technical Report #549, Department of Statistics, U. C. Berkeley. http://www.stat.berkeley.edu/users/census/549.pdf
Freedman, D. (2002). On specifying graphical models for causation, and the identification problem. Technical Report #601, Department of Statistics, U. C. Berkeley. http://www.stat.berkeley.edu/users/census/601.pdf
Freedman, D., Gold, L. S. and Lin, T. H. (1996). Concordance between rats and mice in bioassays for carcinogenesis. *Reg. Tox. Pharmacol.*, 23, 225-232.
Freedman, D. and Navidi, W. (1989). On the multistage model for carcinogenesis. *Environ. Health Perspect.*, 81, 169-188.
Freedman, D. and Navidi, W. (1990). Ex-smokers and the multistage model for lung cancer. *Epidemiol.*, 1, 21-29.
Freedman, D. and Petitti, D. (2001). Salt and blood pressure: Conventional wisdom reconsidered. *Evaluation Review*, 25, 267-287.
Freedman, D., Pisani, R., and Purves, R. (1997). *Statistics.* 3rd ed. Norton, New York.
Freedman, D. and Zeisel, H. (1988). From mouse to man: the quantitative assessment of cancer risks. *Statistical Science*, 3, 3-56 (with discussion).
Friedman, M. (1953). *Essays in Positive Economics.* University of Chicago Press.
Gagnon, F. (1950). Contribution to the study of the etiology and prevention of cancer of the cervix. *Amer. J. Obstetrics and Gynecology*, 60, 516-522.
Gail, M. H. (1996). Statistics in action. *J. Amer. Statist. Assoc.*, 433, 1-13.
Gamble, J. F. (1998). PM_{25} and mortality in long-term prospective cohort studies: cause-effect or statistical associations? *Environ. Health Perspect.*, 106, 535-549.

Gardner, M. J. (1992). Leukemia in children and paternal radiation exposure at the Sellafield nuclear site. *Mon. Nat. Cancer Inst.,* 12, 133-135.
Gardner, M. J., Snee, M. P., Hall, A. J., Powell, C. A., Downes, S. and Terrell, J. D. (1990). Results of case-control study of leukaemia and lymphoma among young people near Sellafield nuclear plant in West Cumbria. *Br. Med. J.,* 300, 423-433. Published erratum appears in BMJ, 1992, 305 715, and see letter in BMJ, 1991, 302 907.
Gauss, C. F. (1809). *Theoria Motus Corporum Coelestium.* Perthes et Besser, Hamburg. Reprinted in 1963 by Dover, New York.
Gavarret, J. (1840). *Principes généraux de statistique médicale, ou, Développement des règles qui doivent présider à son emploi.* Bechet jeune et Labe, Paris.
Goldthorpe, J. H. (1998). *Causation, Statistics and Sociology.* Twenty-ninth Geary Lecture, Nuffield College, Oxford. Publ. by the Economic and Social Research Institute, Dublin, Ireland.
Goldthorpe, J. H. (2001). Causation, statistics, and sociology. *European Sociological Review,* 17, 1-20.
Greenland, S., Pearl, J., and Robins, J. M. (1999). Causal diagrams for epidemiologic research. *Epidemiol.,* 10, 37-48.
Hakama, M., Lehtinen, M., Knekt, P., Aromaa, A., Leinikki, P., Miettinen, A., Paavonen, J., Peto, R. and Teppo, L. (1993). Serum antibodies and subsequent cervical neoplasms: A prospective study with 12 years of followup. *Amer. J. Epidemiol.,* 137, 166-170.
Heckman, J. J. (2000). Causal parameters and policy analysis in economics: A twentieth century retrospective. *The Quarterly Journal of Economics,* CVX, 45-97.
Hodges, J. L. and Lehmann, E. L. (1964). *Basic Concepts of Probability and Statistics.* Holden-Day, San Francisco.
Holland, P. (1988). Causal inference, path analysis, and recursive structural equations models. In C. Clogg, (ed.), *Sociological Methodology,* 449-493.
Howard-Jones, N. (1975). *The Scientific Background of the International Sanitary Conferences 1851-1938.* World Health Organization, Geneva.
Humphreys, P. and Freedman, D. (1996). The grand leap. *Brit. J. Phil. Sci,.* 47, 113-123.
Humphreys, P. and Freedman, D. (1999). Are there algorithms that discover causal structure? *Synthese,* 121, 29-54.
IARC (1986). *Tobacco Smoking.* International Agency for Research on Cancer, Monograph 38, Lyon. Distributed by Oxford University Press.
Kanarek, M. S., Conforti, P. M., Jackson, L. A., Cooper, R. C., and Murchio, J. C. (1980). Asbestos in drinking water and cancer incidence in the San Francisco Bay Area. *Amer. J. Epidemiol.,* 112, 54-72.
Kaprio, J. and Koskenvuo, M. (1989). Twins, smoking and mortality: a 12-year prospective study of smoking-discordant twin pairs. *Social Science and Medicine,* 29, 1083-1089.
Kinlen, L. J. and John, S. M. (1994). Wartime evacuation and mortality from childhood leukaemia in England and Wales in 1945-49. *Br. Med. J.,* 309, 1197-1201.
Lang, J. M., Rothman, K. J., Cann, C. I. (1998). That confounded *P*-value. *Epidemiology,* 9, 7-8.
Legendre, A. M. (1805). *Nouvelles méthodes pour la détermination des orbites des comètes.* Courcier, Paris. Reprinted in 1959 by Dover, New York.
Lieberson, S. (1985). *Making it Count.* University of California Press, Berkeley.
Liu, T. C. (1960). Under-identification, structural estimation, and forecasting. *Econometrica,* 28, 855-865.
Lombard, H. L. and Doering, C. R. (1928). Cancer studies in Massachusetts: Habits, characteristics and environment of individuals with and without lung cancer. *New Engl. J. Med.,* 198, 481-487.
Louis, P. (1835). *Recherches sur les effets de la saignée dans quelques maladies inflammatoires: et sur l'action de l'émétique et des vésicatoires dans la pneumonie.* J. B. Baillière, Paris. Reprinted by The Classics of Medicine Library, Birmingham, Alabama, 1986.
Lucas, R. E. Jr. (1976). Econometric policy evaluation: a critique. In K. Brunner and A. Meltzer (eds.), *The Phillips Curve and Labor Markets,* vol. 1 of the Carnegie-Rochester Conferences on Public Policy, supplementary series to the *Journal of Monetary Economics,* North-Holland, Amsterdam, pp. 19-64. (With discussion.)

Manski, C. F. (1995). *Identification Problems in the Social Sciences.* Harvard University Press. Cambridge.
McKim, V. and Turner, S., eds. (1997). *Causality in Crisis? Proceedings of the Notre Dame Conference on Causality,* Notre Dame Press.
Mill, J. S. (1843). *A System of Logic, Ratiocinative and Inductive.* John W. Parker, London. 8th ed. reprinted by Longman, Green and Co., Ltd., London (1965). See especially Book III Chapter VIII. Reprinted in 1974 by the University of Toronto Press.
Miller, J. F., Mekalanos, J. J. and Fialkow, S. (1989). Coordinate regulation and sensory transduction in the control of bacterial virulence. *Science,* 243, 916-922.
Müller, F. H. (1939). Tabakmissbrauch und Lungcarcinom. *Zeitschrift fur Krebsforsuch,* 49, 57-84.
National Research Council (1997). *Possible Health Effects of Exposure to Residential Electric and Magnetic Fields.* National Academy of Science, Washington, DC.
Neyman, J. (1923). Sur les applications de la théorie des probabilités aux experiences agricoles: Essai des principes. *Roczniki Nauk Rolniczki,* 10, 1-51, in Polish. English translation by D. Dabrowska and T. Speed, 1990. *Statistical Science,* 5, 463-480.
Ni Bhrolcháin, M. (2001). Divorce effects and causality in the social sciences. *European Sociological Review,* 17, 33-57.
Ottenbacher, K. J. (1998). Quantitative evaluation of multiplicity in epidemiology and public health research. *Amer. J. Epidemiol.,* 147, 615-619.
Paneth, N., Vinten-Johansen, P., Brody, H. and Rip, M. (1998). A rivalry of foulness: official and unofficial investigations of the London cholera epidemic of 1854. *Amer. J. Publ. Health,* 88, 1545-1553.
Pasteur, L. (1878). *La théorie des germes et ses applications à la médecine et à la chirurgie,* lecture faite à l'Academie de Médecine le 30 avril 1878, par M. Pasteur en son nom et au nom de MM. Joubert et Chamberland, G. Masson, Paris.
Pearl, J. (1995). Causal diagrams for empirical research. *Biometrika,* 82, 689-709.
Pearl, J. (2000). *Causality: Models, Reasoning, and Inference.* Cambridge University Press. Cambridge.
Perneger, T. V. (1998). What's wrong with Bonferroni adjustments? *Br. Med. J.,* 316, 1236-1238.
Pope, C. A., Schwartz, J. and Ransom, M. R. (1992). Daily mortality and PM_{10} pollution in Utah Valley. *Archives of Environmental Health,* 47, 211-217.
Quetelet, A. (1835). *Sur l'homme et le développement de ses facultés, ou Essai de physique sociale.* Bachelier, Paris
Raufman, J. P. (1998). Cholera. *Amer. J. Med.,* 104, 386-394.
Robinson, W. S. (1950). Ecological correlations and the behavior of individuals. *Amer. Sociol. Rev.,* 15, 351-357.
Rojel, J. (1953). *The Interrelation between Uterine Cancer and Syphilis.* Copenhagen.
Rosenberg, C. E. (1962). *The Cholera Years.* Chicago University Press.
Rothman, K. J. (1990). No adjustments are needed for multiple comparisons. *Epidemiol.,* 1, 43-46.
Rothman, K. J. (1996). Lessons from John Graunt. *Lancet,* 347, 37-39.
Rothman, K. J. and Greenland, S., eds. (1998). *Modern Epidemiology,* 2nd. ed. Lippincott-Raven. Philadelphia.
Rubin, D. (1974). Estimating causal effects of treatments in randomized and nonrandomized studies. *Journal of Educational Psychology,* 66, 688-701.
Scharfstein, D. O., Rotnitzky, A., and Robins, J. M. (1999). Adjusting for non-ignorable drop-out using semiparametric non-response models. *J. Amer. Statist. Assoc.,* 94, 1096-1146 (with discussion).
Semmelweiss, I. (1867). *The Etiology, Concept, and Prophylaxis of Childbed Fever.* Translated by K. C. Carter, University of Wisconsin Press, 1983.
Snow, J. (1855). *On the Mode of Communication of Cholera.* Churchill, London. Reprinted by Hafner, New York, 1965.
Stigler, S. M. (1986). *The History of Statistics.* Harvard University Press. Cambridge, Massachusetts.
Stolley, P. (1991). When genius errs. *Amer. J. Epidemiol.,* 133, 416-425.

Storey, A., Thomas, M, Kalita, A., Harwood, C., Gardiol, D., Mantovani, F., Breuer, J., Leigh, I. M., Matlashewski, G. and Banks, L. (1998). Role of a p53 polymorphism in the development of human papillomavirus-associated cancer. *Nature,* 393, 229-234.

Styer, P., McMillan, N., Gao, F., Davis, J. and Sacks, J. (1995). Effect of outdoor airborne particulate matter on daily death counts. *Environ. Health Perspect.,* 103, 490-497.

Taubes, G. (1995). Epidemiology faces its limits. *Science,* 269, (14 July) 164-169. Letters (8 Sep) 1325-1328.

Taubes, G. (1998). The (political) science of salt. *Science,* 281, (14 August) 898-907.

Terris, M., ed. (1964). *Goldberger on Pellagra.* Louisiana State University Press. Baton Rouge.

Vandenbroucke, J. P. and Pardoel, V. P. (1989). An autopsy of epidemiologic methods: the case of 'poppers' in the early epidemic of the acquired immunodeficiency syndrome (AIDS). *Amer. J. Epidemiol.,* 129, 455-457; and see comments.

Wald, N. and Nicolaides-Bouman, A., eds. (1991). *UK Smoking Statistics.* 2nd ed., Oxford University Press. Oxford.

Winkelstein, W. (1995). A new perspective on John Snow's communicable disease theory. *Amer. J. Epidemiol.,* 142, (9 Suppl.) S3-9.

Wynder, E. L. and Graham, E. A. (1950). Tobacco smoking as a possible etiological factor in bronchogenic carcinoma: a study of six hundred and eight-four proved cases. *J. Amer. Med. Assoc.,* 143, 329-336

Wynder, E. L., Cornfield, J., Schroff, P. D. and Doraiswami, K. R. (1954). A study of environmental factors in carcinoma of the cervix. *American Journal of Obstetrics and Gynecology,* 68, 1016-1052.

Yule, G. U. (1899). An investigation into the causes of changes in pauperism in England, chiefly during the last two intercensal decades. *J. Roy. Statist. Soc.,* 62, 249-295.

Received: January 2002

6
SCANNING A LATTICE FOR A PARTICULAR PATTERN

J. Gani
Mathematical Sciences Institute
Australian National University, Canberra, Australia

1. Introduction

In some earlier work, Gani (1998a, 1998b) and Gani and Irle (1999) have examined problems of patterns arising in sequences of length $n > 0$ of random events, whether independent or Markovian. Here, we apply the known results to the case where these events form patterns on a $2 \times n$ lattice such as that illustrated in Figure 6.1., where the (a_j, b_j, c_j, d_j) are elements selected from some alphabet of size M.

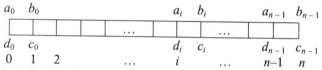

Figure 6.1. A $2 \times n$ lattice

A possible interpretation, though not the only one, is that the lattice forms a strand of DNA, with the horizontal lines in fact coiled helically around each other, and one of the $M = 4$ nucleotide bases A, C, G, T located at each of the positions (a_j, b_j, c_j, d_j), $j = 0, 1, ..., n - 1$, with A and T, or C and G at opposite ends. Usually for a gene, the strand terminates with a particular repetition or pattern at the positions $(a_{n-2}, b_{n-2}, c_{n-2}, d_{n-2})$, $(a_{n-1}, b_{n-1}, c_{n-1}, d_{n-1})$. The problem is to determine the length N of the strand for the first occurrence of this pattern.

To simplify the framework of the problem, and highlight the principles involved, we shall consider an alphabet of $M = 2$ events only, denoting these by 0 and 1. There are then 16 possible patterns (a_j, b_j, c_j, d_j) for the events which may be denoted by

$$\begin{array}{lllll}
A_1 : 0,0,0,0 & A_2 : 1,0,0,0 & A_6 : 1,1,0,0 & A_{12} : 1,1,1,0 & A_{16} : 1,1,1,1 \\
& A_3 : 0,1,0,0 & A_7 : 1,0,1,0 & A_{13} : 1,1,0,1 & \\
& A_4 : 0,0,1,0 & A_8 : 1,0,0,1 & A_{14} : 1,0,1,1 & \\
& A_5 : 0,0,0,1 & A_9 : 0,1,0,1 & A_{15} : 0,1,1,1 & \quad (1.1) \\
& & A_{10} : 0,0,1,1 & & \\
& & A_{11} : 0,1,1,0 & &
\end{array}$$

Let us assume that these events form a Markov chain, so that at the ith step,

$$P\{A_k^{(i)} \mid A_j^{(i-1)}\} = p(A_j A_k) \tag{1.2}$$

where $A_j^{(i-1)} = (a_{i-1}, b_{i-1}, c_{i-1}, d_{i-1})$, and $A_k^{(i)} = (a_i, b_i, c_i, d_i)$, but with $b_{i-1} = a_i$ and $c_{i-1} = d_i$. If we examine the patterns $A_1, \ldots, A_{16}$, we see that

$$\begin{array}{rll}
A_1, A_2, A_5, A_8 & \text{can be followed only by} & A_1, A_3, A_4, A_{11}, \\
A_3, A_6, A_9, A_{13} & \text{can be followed only by} & A_2, A_6, A_7, A_{12}, \\
A_4, A_7, A_{10}, A_{14} & \text{can be followed only by} & A_5, A_9, A_{10}, A_{15}, \\
A_{11}, A_{12}, A_{15}, A_{16} & \text{can be followed only by} & A_8, A_{13}, A_{14}, A_{16}.
\end{array}$$

The transition probability matrix for this Markovian chain will have the following configuration P of (1.3), where $j = 1, \ldots, 16$ stand for A_j, and the appropriate probabilities (1.2) are placed in the positions (j, k). All initial probabilities $p(A_j)$ are assumed to be known.

$$P = \begin{array}{c|cccccccccccccccc}
\nearrow & 1 & 2 & 3 & 4 & 5 & 6 & 7 & 8 & 9 & 10 & 11 & 12 & 13 & 14 & 15 & 16 \\
\hline
1 & * & & * & * & & & & & & & * & & & & & \\
2 & * & & * & * & & & & & & & * & & & & & \\
3 & & * & & & & * & * & & & & & * & & & & \\
4 & & & & & * & & & & * & * & & & & & * & \\
5 & * & & * & * & & & & & & & * & & & & & \\
6 & & * & & & & * & * & & & & & * & & & & \\
7 & & & & & * & & & & * & * & & & & & * & \\
8 & * & & * & * & & & & & & & * & & & & & \\
9 & & * & & & & * & * & & & & & * & & & & \\
10 & & & & & * & & & & * & * & & & & & * & \\
11 & & & & & & & & * & & & & & * & * & & * \\
12 & & & & & & & & * & & & & & * & * & & * \\
13 & & * & & & & * & * & & & & & * & & & & \\
14 & & & & & * & & & & * & * & & & & & * & \\
15 & & & & & & & & * & & & & & * & * & & * \\
16 & & & & & & & & * & & & & & * & * & & *
\end{array} \tag{1.3}$$

This is a sparse matrix, which can readily be handled numerically; this does not, however, provide much insight into the problem. We prefer to simplify it further, so as to derive some analytic results illustrating our methods.

2. Simplifying Assumptions

The probability $p(A_j A_k)$ of (1.2) could in some cases be simplified, so that

$$p(A_j A_k) = P(B_k^{(i)} \mid B_j^{(i-1)}) \tag{2.1}$$

where $B_k^{(i)} = (b_i, c_i)$, and we now need to deal with only four states, $j, k = 1, 2, 3, 4$, given by

$$B_1 : 0, 0 \qquad B_2 : 0, 1 \qquad B_3 : 1, 0 \qquad B_4 : 1, 1.$$

In this case, the transition probability matrix will have the form

$$\begin{array}{c|cccc} \searrow & B_1 & B_2 & B_3 & B_4 \\ \hline B_1 & p_{11} & p_{12} & p_{13} & p_{14} \\ B_2 & p_{21} & p_{22} & p_{23} & p_{24} \\ B_3 & p_{31} & p_{32} & p_{33} & p_{34} \\ B_4 & p_{41} & p_{42} & p_{43} & p_{44} \end{array} \tag{2.2}$$

and the initial probabilities may be written

$$P(B_1) = p_1, \qquad P(B_2) = p_2, \qquad P(B_3) = p_3, \qquad P(B_4) = p_4.$$

Thus if we required $p(A_1 A_4)$, this would be given by $p_1 p_{11} p_{12}$, while $p(A_7 A_5)$, say, would be $p_3 p_{32} p_{21}$, and so on.

Let us assume for simplicity that the strand will be completed when patterns $B_2 B_4$ or $B_3 B_4$ appear, that is when the following arrays occur in the lattice

$$\begin{array}{cc} 0 & 1 \\ \hline & \\ \hline 1 & 1 \end{array} \qquad \begin{array}{cc} 1 & 1 \\ \hline & \\ \hline 0 & 1 \end{array}$$

In fact, genes in DNA strands usually terminate with a repetition of three nucleotide bases, but the principle involved is the same. As in Gani and Irle (1999), we may construct the augmented transition probability matrix for the original states B_1, B_2, B_3, B_4, and the terminal states $B_2 B_4$ and $B_3 B_4$, so that

$$\begin{array}{c|cccc:cc} \searrow & B_1 & B_2 & B_3 & B_4 & B_2 B_4 & B_3 B_4 \\ \hline B_1 & p_{11} & p_{12} & p_{13} & p_{14} & 0 & 0 \\ B_2 & p_{21} & p_{22} & p_{23} & 0 & p_{24} & 0 \\ B_3 & p_{31} & p_{32} & p_{33} & 0 & 0 & p_{34} \\ B_4 & p_{41} & p_{42} & p_{43} & p_{44} & 0 & 0 \\ \hdashline B_2 B_4 & 0 & 0 & 0 & 0 & 1 & 0 \\ B_3 B_4 & 0 & 0 & 0 & 0 & 0 & 1 \end{array} = \left[\begin{array}{c|c} P & Q \\ \hline 0 & I \end{array} \right]. \tag{2.3}$$

It follows from this that the probability of the strand length is

$$P\{N=n\} = p' P^{n-2} QE, \quad n \geq 2, \tag{2.4}$$

where $p' = [p_1, p_2, p_3, p_4]$, and $E' = [1, 1]$. The generating function of this probability is

$$f(s) = \sum_{n=2}^{\infty} s^2 p' (Ps)^{n-2} QE = s^2 p' (I - Ps)^{-1} QE. \tag{2.5}$$

If the matrix P can be written in its canonical form as DVD^{-1}, where V is the diagonal matrix of its eigenvalues v_1, v_2, v_3, v_4, then

$$(I - Ps)^{-1} = D(I - Vs)^{-1} D^{-1}$$

$$= D \begin{bmatrix} (1-v_1 s)^{-1} & 0 & 0 & 0 \\ 0 & (1-v_1 s)^{-1} & 0 & 0 \\ 0 & 0 & (1-v_1 s)^{-1} & 0 \\ 0 & 0 & 0 & (1-v_1 s)^{-1} \end{bmatrix} D^{-1}, \tag{2.6}$$

and (2.5) can be simplified. The mean length of the strands is $f'(1)$, and its variance is $f''(1) + f'(1) - [f'(1)]^2$. We now consider a particular example.

3. A Special Case

Suppose that in a particular example of (2.2), the transition probabilities are

$$\begin{bmatrix} 1-p & p/2 & p/2 & 0 \\ p/2 & 1-p & 0 & p/2 \\ p/2 & 0 & 1-p & p/2 \\ 0 & p/2 & p/2 & 1-p \end{bmatrix} \tag{3.1}$$

We then see that in the augmented matrix (2.3), P and Q are as follows:

$$P = \begin{bmatrix} 1-p & p/2 & p/2 & 0 \\ p/2 & 1-p & 0 & 0 \\ p/2 & 0 & 1-p & 0 \\ 0 & p/2 & p/2 & 1-p \end{bmatrix}, \quad Q = \begin{bmatrix} 0 & 0 \\ p/2 & 0 \\ 0 & p/2 \\ 0 & 0 \end{bmatrix}. \tag{3.2}$$

The canonical form of P is given by

$$\begin{bmatrix} 1 & 1 & 0 & 0 \\ r & -r & -1 & 0 \\ r & -r & 1 & 0 \\ 1 & 1 & 0 & 1 \end{bmatrix} \begin{bmatrix} 1-p(1-r) & 0 & 0 & 0 \\ 0 & 1-p(1+r) & 0 & 0 \\ 0 & 0 & 1-p & 0 \\ 0 & 0 & 0 & 1-p \end{bmatrix} \begin{bmatrix} 1/2 & r/2 & r/2 & 0 \\ 1/2 & -r/2 & -r/2 & 0 \\ 0 & -1/2 & 1/2 & 0 \\ -1 & 0 & 0 & 1 \end{bmatrix}, \quad (3.3)$$

where $r = 2^{-1/2}$. It follows that

$$QE = \begin{bmatrix} pr/2 \\ -pr/2 \\ 0 \\ 0 \end{bmatrix},$$

and hence the p.g.f. $f(s)$ is given simply by

$$f(s) = \frac{s^2 pr}{2} \left(\frac{p_1 + p_2 r + p_3 r + p_4}{1 - [1 - p(1-r)]s} - \frac{p_1 - p_2 r - p_3 r + p_4}{1 - [1 - p(1+r)]s} \right) \quad (3.4)$$

If $p_1 = p_2 = p_3 = p_4 = 0.25$, and $p = 0.4$, then straightforward calculation leads to the p.g.f.

$$f(s) = s^2 \left(\frac{0.120710677}{1 - 0.882842712\, s} - \frac{0.020710678}{1 - 0.317157287\, s} \right), \quad (3.5)$$

from which the mean $f'(1) = 9.7499$, and $\mathrm{Var}(N) = 64.6875$ so that the standard deviation of N is 8.0429, not too different from the mean. Thus, it is possible for the strand to be completed minimally in 2 steps, or alternatively only after a long sequence.

4. Coding Errors

Whenever a sequence of the type just described is produced, there is a chance, however small, that a coding error may occur. In a DNA strand, this may cause mutation. We examine the case in which the simplified model described in Section 2 is subject to such coding errors.

Suppose that the probability of a coding error in the sequence is $0 < a < 1$, where a is small. We may ask what the probability of $r \leq n - 2$ such errors in a strand of length n may be. We shall write the probability of r errors in a strand of length n as the binomial

$$P\{r \text{ errors in an } n\text{-strand}\} = \frac{(n-2)!}{(n-2-r)!\,r!}\, a^r (1-a)^{n-2-r}. \qquad (4.1)$$

Then, from (2.4), if we sum over all strands of length $r \geq n-2$, we find that the probability of r coding errors irrespective of strand length is

$$P\{r \text{ errors}\} = \sum_{n-2=r}^{\infty} p'\, P^{n-2} QE\, \frac{(n-2)!}{(n-2-r)!\,r!}\, a^r (1-a)^{n-2-r} \qquad (4.2)$$
$$= p'\,(I - P(1-a))^{-(r+1)} (aP)^r QE.$$

We can see directly that the probability generating function of the number of errors will be

$$g(s) = p'\,(I - P(1-a))^{-1} [I - (I - P(1-a))^{-1} aPs]^{-1} QE. \qquad (4.3)$$

There are many interesting problems to resolve in this area; it is my hope that this elementary description of a few of these will encourage others to tackle them.

5. References

Gani, J. (1998a). 'On sequences of events with repetitions.' *Stoch. Models* 14, 265–271.
Gani, J. (1998b). 'Elementary methods for failure due to a sequence of Markovian events.' *J. Appl. Maths. Stoch. Anal.* 11, 311–318.
Gani, J. and Irle, A. (1999). 'On patterns in sequences of random events.' *Monatsh. f. Math.* 127, 295–309.

Received: March 2000, Revised: July 2002

7
MIXTURES EVERYWHERE

Dimitris Karlis and Evdokia Xekalaki
Department of Statistics
Athens University of Economics & Business, Greece

1. Introduction to Mixture Models

Mixture models are widely used in statistical modeling since they can model situations which a simple model cannot adequately describe. In recent years, mixture modeling has been exploited mainly due to high-speed computers that can make tractable problems that occur when working with mixtures (e.g. estimation). Statistics has benefited immensely by the development of advanced computer machines and thus more sophisticated and complicated methodologies have been developed. Mixture models underlie the use of such methodologies in a wide spectrum of practical situations where the hypothesized models can be given a mixture interpretation as demonstrated in the sequel.

In general, mixtures provide generalizations of simple models. For example, assuming a specific form for the distribution of the population that generated a data set implies that the mean to variance relation is given for this distribution. In practical situations this may not always be true. A simple example is the Poisson distribution. It is well known (see, e.g., Johnson et al., 1992) that for the Poisson distribution the variance is equal to the mean. Hence, assuming a Poisson distribution implies a mean to variance ratio equal to unity. With real data sets however, this is rarely the case. Quite often, the sample mean is noticeably exceeded by the sample variance. This situation is known as *overdispersion*. A Poisson distribution is no longer a suitable model in such a case and the need of a more general family of distributions becomes obvious. Such a flexible family may be defined if one allows the parameter (or the parameters) θ of the original distribution to vary according to a distribution with probability density function, say $g(\cdot)$.

Definition 1. A distribution function $F(\cdot)$ is called *a mixture of the distribution function* $F(\cdot \mid \theta)$ *with mixing distribution* $G_\theta(\cdot)$ if it can be written in the form

$$F_x(x) = \int_\Theta F_{x\mid\theta}(x \mid \theta) dG_\theta(\theta), \qquad (1)$$

where Θ is the space in which θ takes values and $G_\theta(\cdot)$ can be continuous, discrete or a finite step distribution.

The above definition can be also expressed in terms of probability density functions in the continuous case (or the probability functions in the discrete case). The above mixture is denoted as $F_{x|\theta}(x) \underset{\theta}{\wedge} G(\theta)$. In the sequel, a mixture with a finite step mixing distribution will be termed *a k-finite step mixture* of $F(\cdot | \theta)$, where k is a non-negative integer referring to the number of points with positive probabilities in the mixing distribution.

Mixture models cover several distinct fields of the statistical science. Their broad acceptance as plausible models in diverse situations is reflected in the statistical literature. Titterington et al. (1985) provide an extensive review of work in the area of mixture models up to 1985. In recent years, the number of applications increased mainly because of the availability of high speed computer resources. Moreover, since many methods can be seen through the prism of mixture models, there is a vast literature concerning applications of mixture models in various contexts. Recent reviews on mixtures can be found in Lindsay (1995), Bohning (1999), McLachlan and Peel (2001).

The purpose of this paper is to bring together various models from diverse fields that are in fact mixture models. The resulting collection of models may be far from being exhaustive as the focus has been on methodologies that are common in statistical practice and not on results concerning special cases. To this extent, the number of articles cited was kept to a minimum and reference was made only to a selection of papers that could pilot the reader in the various areas.

In Section 2 of the chapter two basic concepts are discussed in the context of which the mixture models are used: overdispersion and inhomogeneity. Section 3 presents various statistical methodologies that use the idea of mixtures. An attempt is made to show clearly the connection of such methodologies to mixture models. Finally, in Section 4 a brief discussion is provided highlighting the implications of a unified treatment of all the models discussed.

2. General Properties

2.1 Inhomogeneity models

Mixture models are used to describe inhomogeneous populations. The i-th group of individuals of the population has a distribution defined by a probability density function $f(\cdot | \theta_i)$. All the members of the population follow the same parametric form of distribution, but the parameter θ_i varies from individual to individual according to a distribution $G_\theta(\cdot)$. For example, considering the number of accidents incurred by a population of clients of an insurance company, it is reasonable to assume that there are at least two subpopulations, the new drivers and the old drivers. Drivers can thus be assumed to incur accidents at rates that differ from one subpopulation to the other subpopulation, say $\theta_1 \neq \theta_2$. This is the simplest form of inhomogeneity: the population consists of two subpopulations. Allowing for the number of subpopulations to tend to infinity, i.e., considering different categories of drivers according to infinitely many characteristics, such as age, sex, origin, social, and economic status, etc. a continuous mixing distribution for the parameter θ of the Poisson distribution arises.

Depending on the choice of the mixing distribution $G_\theta(\cdot)$, a very broad family of distributions is obtained, which may be adequate for cases where the simple model fails. So, a mixture model describes an inhomogeneous population while the mixing distribution describes the inhomogeneity of the population. If the population were homogeneous, then all the members would have the same parameter θ, and the simple model would adequately describe the situation.

2.2 Overdispersion

A fundamental property of mixture models stems from the following representation of the variance of the mixed variate X.

$$\mathrm{Var}(X) = \mathrm{Var}(E(X \mid \theta)) + E(\mathrm{Var}(X \mid \theta)). \qquad (2)$$

The above formula separates the variance of X into two parts. Since the parameter θ represents the inhomogeneity of the population, the first part of the variance represents the variance due to the variability of the parameter θ, while the second part reflects the inherent variability of the random variable X if θ did not vary. One can recognize that a similar idea is the basis for ANOVA models where the total variability is split into the "between groups" and the "within groups" components. This is further discussed in Section 3.

The above formula offers an explanation as to why mixture models are often termed as *overdispersion models*. A mixture model has a variance greater than that of the simple model (e.g., Shaked, 1980). Thus, it is commonly proposed that if the simple model cannot describe the variability present in the data, overdispersed alternatives based on mixtures could be used.

3. Fields of Application

3.1 Data Modelling

The main advantage of mixture models lies in that they provide the possibility of generalizing existing simple models through an appropriate choice of a mixing distribution which acts as a means of "loosening" the structure of the initial model by allowing its parameter to vary. A wealth of alternative models can thus be considered whenever the simple (initial) model fails and many interesting distributions may be obtained from simple and well-known distributions such as the Poisson, the binomial, the normal, the exponential, through mixing procedures.

7. MIXTURES EVERYWHERE

In recent years, the computational difficulties for applying such complicated models have disappeared and some new distributions (discrete or continuous) have been proposed. Moreover, since mixture models are widely used to describe inhomogeneous populations they have become a very popular choice in practice, since they offer realistic interpretations of the mechanisms that generated the data.

The derivation of the negative binomial distribution, as a mixture of the Poisson distribution with a gamma distribution as the mixing distribution, originally obtained by Greenwood and Yule (1920) constitutes a typical example. Almost all the well-known distributions have been generalized by considering mixtures of them. A large number of Poisson mixtures have been developed. (For an extensive review, see Karlis, 1998).

Perhaps, the beta binomial distribution (see, e.g., Tripathi et al., 1994) is the most famous example of binomial mixtures. Alternative models have been described in Alanko and Duffy (1996) and Brooks et al. (1997).

Negative binomial mixtures have also been widely used with applications in a large number of fields. These include the Yule distribution (Yule, 1925, Simon, 1955, Kendall, 1961, Xekalaki, 1983a, 1984b) and the generalized Waring distribution (Irwin, 1963, 1968, 1975, Dacey, 1972, Xekalaki, 1983b, 1984a). Note that negative binomial mixtures can be seen as Poisson mixtures as well.

Normal mixtures on the parameter representing the mean of the distribution are not common in practice. Mixtures of the normal distribution on the parameter representing its variance are referred to as *scale mixtures* (e.g., Andrews and Mallows, 1974). For example, the t-distribution is a scale mixture of the normal distribution with a chi-square mixing distribution. Barndorff-Nielsen et al. (1982) described a more general family of normal mixtures of the form $f_{N(\mu+\theta\beta,\theta a)} \wedge_\theta g(\theta)$, where $f_{N(\alpha,\beta)}$ stands for the probability density function of the normal distribution with mean α and variance β. The distributions arising from such mixtures are not necessarily symmetric and have heavier tails than the normal distribution. Applications of normal scale mixtures have been considered by Barndorff-Nielsen (1997) and Eberlein and Keller (1995).

Similarly, exponential mixtures are described in Hebert (1994) and Jewell (1982), for life testing applications. The beta distribution can be seen as a Gamma mixture, while the Gamma distribution can be seen as a scale mixture of the exponential distribution (Gleser, 1989). Many other mixture distributions have been proposed in the literature.

A wide family of distributions can be defined to consist of finite mixtures of distributions, with components not necessarily from the same family of distributions. Finite mixtures with different component distributions have been described in Rachev and Sengupta (1993) (Laplace - Weibull), Jorgensen et al. (1991) (Inverse Gaussian − Reciprocal Inverse Gaussian), Scallan (1992) (Normal − Laplace), Al-Hussaini and Abd-El-Hakim (1989) (Inverse Gaussian-Weibull) and many others.

Finally, note that mixture models can have a variety of shapes that are never taken by simple models, such as multimodal shapes. These are usually represented via finite mixtures. So, for example, mixing two normal distributions of equal variances in equal proportions can result in a bimodal distribution with well-separated modes, appropriate for describing data exhibiting such a behavior.

3.2 Discriminant Analysis

In discriminant analysis, one needs to construct rules so as to be able to distinguish the subpopulation from which a new observation comes. Assuming a finite mixture model one may obtain the parameters of the subpopulations from a training set, and then classify the new observations via simple probabilistic arguments (see, e.g., McLachlan, 1992). This approach is also referred to as *statistical pattern recognition* in computer science applications.

Consider a population consisting of k subpopulations, each distributed according to a distribution defined by a density function $f_j(\cdot|\theta_j)$, $j=1,2,\ldots,k$. Suppose further that the size of each subpopulation is p_j. Usually, data used in discriminant analysis also contain variables Z_j, $j=1,2,\ldots,k$, which take the value 1 if the observation belongs to the j-th subpopulation and 0 otherwise. These data are used for estimating the parameters θ_j, p_j and are referred to as *training data*. Then, a new observation x is allocated to each group according to its posterior probability of belonging to the j-th group

$$P(Z_j = 1 | x) = \frac{p_j f_j(x|\theta_j)}{\sum_{j=1}^{k} p_j f_j(x|\theta_j)}.$$

One can recognize the mixture formulation in the above formula, as well as the fact that this formulation comprises the E-step of the EM algorithm for estimation in finite mixture models. The variables Z_{ij} are the *"missing"* data in the construction of the EM algorithm for finite mixtures.

However, such data sets often contain a lot of unclassified observations, i.e., observations that do not relate to specific values of Z_j, $j=1,2,\ldots,k$ and hence one can use these data for estimation purposes. The likelihood function for such data is expressed in terms of the mixture $\sum_{j=1}^{k} p_j f_j(x|\theta_j)$ and standard mixture methodologies must be used for estimating the parameters. Note that unclassified observations contribute to the estimation of all the parameters (see, e.g., Hosmer, 1973).

Usually, the densities $f_j(\cdot|\theta_j)$ are assumed multivariate normal with both the mean vector and the variance-covariance matrix being variable.

It is interesting to note that although the EM algorithm for mixtures was introduced quite early by Hasselblad (1969), it did not find wide usage until computer machines became widely available. This typically reflects the impact of computer resources in mixture modeling. The same is true of a wide range of fields that, despite their early development, attracted greater interest only after the generalized use of statistical software. Cluster analysis is another typical example.

3.3 Cluster Analysis

Finite mixtures play an important role to the development of methods in cluster analysis. Two main approaches are used for clustering purposes. The first considers distances between the observations and then clusters the data according to their distances from specific cluster centers. The second approach utilizes a finite mixture model.

The idea is to describe the entire population as a mixture model consisting of several subpopulations (clusters). Then, a methodology could be to fit this finite mixture model and subsequently use the estimated parameters to obtain the posterior probability with which each of the observations belongs to the j-th subpopulation (McLachlan & Basford, 1989). According to a decision criterion, each observation is allocated to a subpopulation, thus creating clusters of data. The problem of choosing the number of clusters that best describe the data, reduces to that of selecting the number of support points for the finite mixture (see, e.g., Karlis & Xekalaki, 1999).

Usually, multivariate normal subpopulations are considered (Banfield & Raftery, 1993 and McLachlan & Basford, 1989). Symons et al. (1983) found clusters of Poisson distributed data for an epidemiological application, while data containing both continuous and discrete variables can be analyzed via multivariate normal densities where thresholds are used for the categorical variables (see, e.g., Everitt & Merette, 1990).

3.4 Outlier-robustness Studies

Outliers in data sets have been modelled by means of mixture models (see, e.g., Aitkin & Wilson, 1980). It is assumed that an outlier comprises a component in a mixture model. More formally, the representation used for the underlying model is

$$(1-p)\ f(\cdot|\theta)\ + p\, g(\cdot),$$

where $f(\cdot|\theta)$ is the true density *contaminated* by a proportion of p observations from a density $g(\cdot)$. Hence, by fitting a mixture model we may investigate the existence of outliers. In robustness studies, the contamination of the data can also be regarded as an additional component of a mixture model.

In addition, for robustness studies with normal populations it is natural to use a t-distribution. Recall that the t-distribution is in fact a scale mixture of the normal distribution. Other scale mixtures have also been proposed for examining robustness of methods for normal populations (e.g., Cao & West, 1996).

Note further, that since mixtures of a distribution tend to this distribution if a degenerate mixing distribution is used, it would be natural to consider the general mixture model as leading to the simple model as the variance of the underlying model decreases.

3.5 Analysis of Variance (ANOVA) Models

The wellknown technique of the analysis of variance is a particular application of mixture models. It is assumed that the mean of the normal distribution of the entire population, varies from subpopulation to subpopulation and the total variance is decomposed with respect to randomness and mixing.

The simple ANOVA model assumes prespecified values for the means of the different components, not allowing them to vary. The case where the means come from a distribution with density $g(\cdot)$ corresponds to the so-called *random effects model* described in the sequel. It is interesting that the simple ANOVA models are based on the inhomogeneity model which allows for subpopulations with different means. The decomposition of the variance given in (2) is the classical ANOVA model separating the total variance into the "between groups" variance and the "within groups" variance.

Beyond the widely applied classical ANOVA model for normal populations, similar ANOVA models have been proposed for discrete data as well. For example, Irwin (1968) and Xekalaki (1983b, 1984a), in the context of accident theory, considered analyzing the total variance into three additive components corresponding to internal and external non-random factors and to random factors. Also, Brooks (1984) described an ANOVA model for beta-binomial data.

3.6 Random Effects Models and Related Models

Consider the classical one-way ANOVA model. It is assumed that the i-th observation of the j-th group, say X_{ij}, follows a $N(\theta_j, \sigma^2)$ distribution, where θ_j is the mean of the j-th group. The simple ANOVA model assumes that the values of θ_j's are prespecified. Random effect models assume that the parameters are not constant but they are realizations from a distribution with density $g(\cdot)$. The resulting marginal density function of the data for each group is of the form

$$f(\mathbf{x}_j) = \int_\Theta \prod_{i=1}^{n_j} f(x_{ij}; \theta, \sigma^2) g(\theta) d\theta ,$$

where n_j is the sample size of the j-th subpopulation. The usual choice for $g(\cdot)$ is the density function of the normal distribution, resulting in normal marginals. This choice was based mainly on its computational tractability, since other choices led to complicated marginal distributions.

Such random effects models have been described for the broad family of Generalized Linear Models. Consider, for example, the Poisson regression case. For simplicity, we consider only a single covariate, say X. A model of this type assumes that the data Y_i, $i=1, 2, \ldots, n$ follow a Poisson distribution with mean λ_i such that

$$\log(\lambda_i) = a + \beta X_i + \varepsilon_i$$

for some constants a, β and with ε_i having a distribution with mean equal to 0 and variance say φ. Now the marginal distribution of the observations y_i is no longer the Poisson distribution, but a mixed Poisson distribution, with mixing distribution clearly depending on the distribution of ε_i.

From the regression equation, one can obtain that

$$Y_i \sim \text{Poisson}(t_i \exp(a + \beta X_i)) \wedge_t g(t),$$

where $t_i = \exp(\varepsilon_i)$ with a distribution that depends on the distribution of ε_i. Negative Binomial and Poisson Inverse Gaussian regression models have been proposed as overdispersed alternatives to the Poisson regression model (Lawless, 1987, Dean et al., 1989). If the distribution of t is a two finite step distribution, the finite Poisson mixture regression model of Wang et al. (1996) results. The similarity of the mixture representation and the random effects one is discussed in Hinde and Demetrio (1998).

The above example from the Poisson distribution can easily be generalized. All the random effect models introduce a mixing distribution for the error which adds one more term to the total variability.

Very similar to the random effect model is the formulation of repeated measurement models, where the added variability is due to the variance induced by the series of observations on the same individual. Starting from a linear regression (in general one can consider any link function suitably linearized) one can add variability by regarding any term of the linear regression model as a random variable. So, allowing the intercept parameter to vary leads to *random coefficient regression models* (see Mallet, 1986). Also, *error-in-variables models* arise by letting the covariates themselves vary. Finally, *random effect models* are obtained if the error term is allowed to vary.

Lee and Nelder (1996) discussed the above models under the general caption of *hierarchical generalized linear models*.

3.7 Kernel Density Estimation

In *kernel density estimation*, the aim is to estimate a probability density function $f(.)$ on the basis of a sample of size n by smoothing the probability mass of $\frac{1}{n}$ placed at each of the observations (x_i) by the empirical distribution function according to a kernel $K_n(\cdot, x_i)$. This is usually a symmetric probability density function of the form $K_n(x, x_i) = K\left(\dfrac{x - x_i}{h}\right)$, $i = 1, 2, ..., n$, where h is a switching parameter which handles the smoothing procedure (see, e.g., Silverman, 1986). Thus, in kernel density estimation a kernel mixture model is considered with equal mixing probabilities. More specifically, the density estimate $\hat{f}_n(.)$ of $f(.)$ at the point x is obtained by

$$\hat{f}_n(x) = \frac{1}{n}\sum_{i=1}^{n} K\left(\frac{x - x_i}{h}\right).$$

One can recognize that the above representation of the kernel estimate is a n-finite mixture of the kernel $K(\cdot,\cdot)$. Though this mixture representation has been recognized (see, e.g., Simonoff, 1996), it has not been exploited in practice. The idea is to use certain kernels in order to obtain estimates with useful properties. For example, data restricted on the positive axis can be estimated using exponential or gamma kernels (see, e.g., Chen, 2000), depending on the shape (J-shaped or bell shaped data). Similarly, discrete data can be estimated via Poisson kernels etc.

Moreover, specific approaches can be used in order to achieve certain smoothing properties. By using such approaches the choice of the smoothing parameter h can be reduced to the choice of the kernel parameters so that the smoothing properties be fulfilled. Wang and Van Ryzin (1979) described such an approach in an empirical Bayesian context for the estimation of a Poisson mean. They proposed estimating the discrete density, given the data $X_1, X_2, ..., X_n$ by

$$\hat{P}(x) = \frac{1}{n}\sum_{i=1}^{n} \frac{\exp(-x_i)x_i^x}{x!}$$

i.e., as a mixture of n Poisson distributions with parameters equal to the observations x_i, i=1, 2, ..., n.

3.8 *Latent Structure Models and Factor Analysis*

In latent structure models it is assumed that beyond the observable random variables there are other unobservable or even non-measurable variables, which influence the situation under investigation. The main assumption in the case of latent structure models is that of conditional independence, i.e., the assumption that for a given value of the unobservable variable the remaining variables are independent. Since inference is based on the unconditional distribution, we obtain, by the law of total probability, a mixture model where the mixing distribution represents the distribution of the unobservable quantity which thus is of special interest in many situations (see, e.g., Everitt, 1984). It is very interesting that many methods proposed for mixture models are applicable to latent variable models (see, e.g., Aitkin et al., 1981).

7. MIXTURES EVERYWHERE

For example, in psychological tests the probability that the i-th person will correctly answer x questions is described as $p(x|\varphi_i)$ where φ_i represents the ability of the i-th person. Additionally, it is assumed that given the ability of each person the scores x are independent. (This is the idea of conditional independence). Since, ability is a rather abstract notion that cannot be measured, the researcher may assume either a parametric form of distribution for its values (e.g., a normal distribution with some parameters) or a finite step distribution (as in the case where φ_i can take only a finite number of different values). This has a common element with the method of factor analysis for continuous variables (Bartholomew, 1980).

A formulation of the problem is the following. Suppose that one observes a set of p-variables, say $\mathbf{x} = (x_1, x_2, \ldots, x_p)$. A latent structure model supposes that these variables are related to a set of q unobservable and perhaps non-measurable variables (e.g., some abstract concepts such as hazard, interest, ability, love, etc.), say $\mathbf{y} = (y_1, y_2, \ldots, y_q)$. For the model to be practically useful, q needs to be much smaller than p. The relationship between $\mathbf{x}$ and $\mathbf{y}$ is stochastic and may be expressed by a conditional probability function $\pi(\mathbf{x}|\mathbf{y})$ being the conditional distribution of the variables $\mathbf{x}$ given the unobservable $\mathbf{y}$. The purpose of latent structure models is to infer on $\mathbf{y}$, keeping in mind, that we have observed only $\mathbf{x}$. The marginal density of $\mathbf{x}$ can be represented as a mixture, by

$$f(\mathbf{x}) = \int \pi(\mathbf{x}|\mathbf{y}) p(\mathbf{y}) d\mathbf{y}.$$

One can infer on $\mathbf{y}$ using the Bayes theorem, since

$$\pi(\mathbf{y}|\mathbf{x}) = \frac{\pi(\mathbf{x}|\mathbf{y}) p(\mathbf{y})}{f(\mathbf{x})}.$$

Hence, the problem reduces to one of estimating the mixing density $p(\cdot)$. As described earlier, this density can be either specified parametrically, and hence only the parameters of the defined density must be estimated, or it can be estimated non-parametrically (see, e.g., Lindsay et al., 1991).

Latent structure models can be considered as factor analysis models for categorical data. The classical factor analysis model assumes that a set of observable variables, say $\mathbf{x} = (x_1, x_2, \ldots, x_p)$ can be expressed as a linear combination of a set of unobservable variables, say $\mathbf{y} = (y_1, y_2, \ldots, y_q)$, termed *factors*. More formally

$$\mathbf{x} = \mathbf{By} + \varepsilon$$

where the matrix $\mathbf{B}$ contains the *factor loadings*, i.e., its (i, j) element is the contribution of the j-th factor to the determination of the i-th variable. The vector of errors ε contains the unexplained part of each variable and it is assumed to follow a $N(0, \mathbf{D})$, where $\mathbf{D} = \text{diag}(\sigma_1^2, \sigma_2^2, \ldots, \sigma_p^2)$. Conditionally on the factors $\mathbf{y}$, $\mathbf{x}|\mathbf{y} \sim N(\mathbf{By}, \mathbf{D})$ distribution, and the factors follow themselves a $N(\mathbf{0}, \mathbf{I}_q)$ distribution. Then, the unconditional distribution of $\mathbf{x}$ is a $N(\mathbf{0}, \mathbf{BB}^t + \mathbf{D})$ distribution. Note that the variance-covariance matrix is decomposed into two terms, the variance explained by the factors and the remaining unexplained part. This decomposition is the basis for the factor analysis model.

3.9 Bayes and Empirical Bayes Estimation.

Bayesian statistical methods have their origin in the well-known Bayes theorem. From a Bayesian perspective, the parameter θ of a density function, say $f(\cdot|\theta)$ has itself a distribution function $g(\cdot)$, termed the *prior*, reflecting one's belief about the parameter and allowing for extra variability. We treat θ as a scalar for simplicity, but clearly, it can be vector valued as well. The prior distribution corresponds to the mixing distribution in (1).

The determination of the prior distribution is crucial for the applicability of the method. Standard Bayesian methods propose a prior based on past experience, on the researcher's belief, or a non-informative prior in the sense that no clear information about the parameter exists and this ignorance is accounted for by a very dispersed prior. Instead of determining the prior by specific values of its parameters, recent hierarchical Bayes models propose treating the parameters of the prior distribution as random variates and imposing hyperpriors on them. Such an approach can remove subjectivity with respect to the selection of the prior distribution.

A different approach is that of the so-called Empirical Bayes methodologies (see, e.g., Karlin & Lewis, 1996). Specifically, the Empirical Bayesian methods aim at estimating the prior distribution from the data. This reduces to the problem of estimating the mixing distribution. This obvious relationship between these two distinct areas of statistics have resulted in a vast number of papers in both areas, with many common elements (see, for example, Maritz & Lwin, 1989, Laird, 1982). The aim is the same in both cases, though the interest lies in different aspects.

Putting aside the relationship of Bayesian approaches and mixture models, there are several other topics in the Bayesian literature that use mixtures in order to improve the inferences made. For example, mixtures have been proposed to be used as priors, the main reason being their flexibility (see, e.g., Dalal & Hall, 1983). Beyond that, such priors are also robust and have been proposed for examining Bayesian robustness (Bose, 1994). Escobar and West (1995) proposed mixtures of normals as an effective basis for nonparametric Bayesian density estimation.

3.10 Random Variate Generation

The mixture representation of some distributions is a powerful tool for efficient random number generation from these distributions. Several distributions (discrete or continuous) may arise as mixture models from certain distributions, which are easier to generate. Hence, generating variables in the context of such a representation can be less expensive.

For example, variables can be generated from the negative binomial distribution by utilizing its derivation as a mixture of the Poisson distribution with a Gamma mixing distribution. Another, more complicated example of perhaps more practical interest is given by Philippe (1997). She considered generation of truncated gamma variables based on a finite mixture representation of the truncated Gamma distribution.

Furthermore, the distributions of products and ratios of random variables can be regarded as mixtures and hence the algorithms used to simulate from such distributions are in fact examples of utilizing their mixture representation. For more details, the reader is referred to Devroye (1992).

3.11 Approximating the Distribution of a Statistic

In many statistical methods, the derived statistics do not have a standard distributional form and an approximation has to be considered for their distribution. Mixture models allow for flexible approximation in such cases. Such an example is the approximation of the distribution of the correlation coefficient used in Mudholkar and Chaubey (1976). In order to cope with the inappropriateness of the normal approximation of the distribution of the sample correlation coefficient, especially in the tails of the distribution, they proposed the use of a mixture of a normal distribution with a logistic distribution. Such a mixture results in a distribution with heavier tails suitable for the distribution of the correlation coefficient

3.12 Multilevel Models

Multilevel statistical models assume that one can separate the total variation of the data into levels and estimate the component attributed to each level (see, e.g., Goldstein, 1995). Consider a k-level model and let (y_{ij}, x_{ij}) denote the i-th observation from the j-th level. In the context of the typical linear model

$$y_{ij} = a_j + b_j x_{ij} + e_{ij}$$

one has to estimate the parameters $a_j, b_j, j=1, \ldots, k$ and the variance σ^2 of the data.

A multilevel statistical model treats the parameters a_j, b_j as random variables in the sense that $a_j = c_0 + u_j$, and $b_j = c_1 + v_j$ where (u_j, v_j) follows a bivariate normal distribution with zero means and a variance covariance matrix. Then, the simple model can be rewritten as

$$y_{ij} = c_o + c_1 x_{ij} + (u_j + v_j x_{ij} + e_{ij}).$$

A variance component is added corresponding to each level. For this reason, the model is also termed as the *variance components model*. Since normal distributions are usually used (mainly for convenience) the resulting distributions are also normal. The mixture representation is used for applying an EM algorithm for the estimation of the parameters (Goldstein, 1995).

3.13 Distributions Arising out of Methods of Ascertainment

When an investigator collects a sample of observations produced by nature according to some model, the original distribution may not be reproduced due to various reasons. These include partial destruction or enhancement of observations. Situations of the former type are known in the literature as *damage models* while situations of the latter type are known as *generating models*. The distortion mechanism is usually assumed to be manifested through the conditional distribution of the resulting random variable Y given the value of the original random variable X. As a result, the observed distribution is a distorted version of the original distribution obtained as a mixture of the distortion mechanism. In particular, in the case of damage,

$$P(Y = r) = \sum_{n=r}^{\infty} P(Y = r | X = n) P(X = n), \qquad r = 0, 1, \ldots,$$

while in the case of enhancement

$$P(Y = r) = \sum_{n=1}^{r} P(Y = r | X = n) P(X = n), \quad r = 1, 2, \ldots$$

Various forms of distributions have been considered for the distortion mechanisms in the above two cases. In the case of damage, the most popular forms have been the binomial distribution (Rao, 1963), mixtures on p of the binomial distribution (e.g., Panaretos, 1982, Xekalaki & Panaretos, 1983) whenever damage can be regarded as additive (Y=X–U, U independent of Y) or in terms of the uniform distribution in (0, x) (e.g., Xekalaki, 1984b) whenever damage can be regarded as multiplicative (Y= [RX], R independent of X and uniformly distributed in (0, 1)). The latter case has also been considered in the context of continuous distributions by Krishnaji (1970b). The generating model was introduced and studied by Panaretos (1983).

3.14 Other Models

It is worth mentioning that several other problems in the statistical literature can be seen through the prism of mixture models. For example, *deconvolution problems* (see, e.g., Caroll & Hall, 1988, Liu & Taylor, 1989) assume that the data X can be written as Y+Z, where Y is a latent variable and Z has a known density f. Then, the density of X can be written in a mixture form, thus

$$g(x) = \int f(x - y) dQ(y),$$

where $Q(\cdot)$ is the distribution function of the latent variable Y. The above model can be considered as a *measurement error model*. In this context, the problem reduces to estimating the mixing distribution from a mixture. Similar problems related to hidden Markov models are described in Leroux (1992).

Simple convolutions can be regarded as mixture models. Also, as already mentioned in Section 3.10, products of random variables can be regarded as mixture models (Sibuya, 1979).

Another application of mixtures is given by Rudas et al. (1994) in the context of testing for the goodness of fit of a model. In particular, they propose treating the model under investigation as a component in a 2-finite mixture model. The estimated mixing proportion together with a parametric bootstrap confidence interval for this quantity can be regarded as evidence for or against the assumed model. The idea can be generalized to a variety of goodness of fit problems, especially for non-nested models.

From this, it becomes evident that a latent mixture structure exists in a variety of statistical models, often ignored by the researcher. Interesting reviews for the mixture models are given in the books by Everitt and Hand (1981), Titterington et al. (1985), McLachlan and Basford (1989), Lindsay (1995), Bohning (1999), McLachlan and Peel (2001) as well as in the review papers of Gupta and Huang (1981), Redner and Walker (1984) and Titterington (1990).

4. Discussion

An anthology of statistical methods and models directly or indirectly related to mixture models was given. In some of them, the mixture idea is often well hidden in the property that a location mixture of the normal distribution is itself a normal distribution. So, the standard normal theory still holds and estimation is not difficult under a normal distribution.

A question that naturally arises is what one can gain by such mixture representations of all these models. As has been demonstrated, beyond the philosophical issue of a unified statistical approach, some elements common in all these models can be brought about. Many of these models have a structure that is of a latent nature such as containing, unobserved quantities that are not measurable, but nevertheless play a key-role in the model.

All the models discussed imply the two basic concepts of inhomogeneity and overdispersion. Further, any mixture model admits an interesting missing data interpretation. Thus, a unifying approach in modeling different situations allows the application of methodologies used in the case of mixtures to other models. Mixture models, for instance, provide the rationale on which the estimation step of the well-known EM-algorithm is based for the estimation of the unknown values of the parameters which are treated as "missing data." For example, Goutis (1993) followed an EM algorithmic approach for a logistic regression model with random effects. In general, such EM algorithms for random effect models can reduce the whole problem to one of fitting a generalized linear model to the simple distribution; such procedures are provided in standard statistical packages. Hence, iterative methods that provide estimates can be constructed. Other techniques can also be applied, like nonparametric estimation. (See, Lindsay, 1995, for an interesting elaboration).

Such approaches reduce the need for specific assumptions when applying the models leading to more widely applicable models.

As mentioned in the introduction, the impact of computer resources on the development of mixture models and on the enhancement of their application potential has been tremendous. Although early work on mixtures relied on computers (e.g., the development of the EM algorithm for finite mixture models by Hasselblad, 1969, and the first attempt for model based clustering by Wolfe, 1970), the progress in this field was rather slow until the beginning of last decade. The implementation of Lindsay's (1983) general maximum likelihood theorem for mixtures, a milestone in mixture modeling, relies on computers. Non-parametric maximum likelihood estimation of a mixing distribution became computationally feasible either via an EM algorithm (as described in detail by McLachlan & Krishnan (1997) and McLachlan & Peel (2001) or via other algorithmic methods a detailed account of which is provided by Bohning (1995). Another example is the development of model-based clustering through mixture models. Such models became accessible by a wide range of research workers from a variety of disciplines, only after the systematic use of computers (see, e.g., McLachlan & Basford, 1989). Finally, Bayesian estimation for mixture models became possible via MCMC methods (see, e.g., Diebolt & Robert, 1994) that required high speed computer resources. The multiplicity of applications of mixtures presented in Section 3, reveals that the problems connected with the implementation of the theoretical results would not have become tractable if it were not for the advancement of computer technology. The results could have remained purely theoretical with a few applications by specialists in certain fields.

5. References

Aitkin, M., Anderson, D. and Hinde, J. (1981). Statistical Modelling of Data on Teaching Styles. *Journal of the Royal Statistical Society, A* 144, 419-461.

Aitkin, M. and Wilson, T. (1980). Mixture Models, Outliers and the EM Algorithm. *Technometrics*, 22, 325-331.

Alanko, T. and Duffy, J.C. (1996). Compound Binomial Distributions for Modeling Consumption Data. *The Statistician*, 45, 269-286.

Al-Husainni E.K. and Abd-El-Hakim, N.S. (1989). Failure Rate of the Inverse Gaussian-Weibull Mixture Model. *Annals of the Institute of Statistical Mathematics*, 41, 617-622.

Andrews, D.F , Mallows, C. L. (1974). Scale Mixtures of Normal Distributions. *Journal of the Royal Statistical Society, B* 36, 99-102.

Banfield, D.J. and Raftery, A.E. (1993). Model-based Gaussian and Non-Gaussian Clustering. *Biometrics*, 49, 803-821.

Barndorff-Nielsen, O.E. (1997). Normal Inverse Gaussian Distributions and Stochastic Volatility Modeling. *Scandinavian Journal of Statistics*, 24, 1-13.

Barndorff-Nielsen, O.E., Kent, J. and Sorensen, M. (1982). Normal Variance-Mean Mixtures and z-Distributions. *International Statistical Review*, 50, 145-159.

Bartholomew, D.J. (1980). Factor Analysis for Categorical Data. *Journal of the Royal Statistical Society*, B, 42, 292-321.

Böhning, D. (1995). A Review of Reliable Maximum Likelihood Algorithms for Semiparametric Mixture Models. *Journal of Statistical Planning and Inference*, 47, 5-28.

Böhning, D. (1999). *Computer Assisted Analysis of Mixtures (C.A.M.AN)*. Marcel Dekker Inc. New York
Bose, S. (1994). Bayeasian Robustness with Mixture Classes of Priors. *Annals of Statistics*, 22, 652-667.
Brooks, R.J. (1984). Approximate Likelihood Ratio Tests in the Analysis of Beta-Binomial Data. *Applied Statistics*, 33, 285-289.
Brooks, S.P., Morgan, B.J.T., Ridout, M.S. and Pack, S.E. (1997). Finite Mixture Models for Proportions. *Biometrics*, 53, 1097-1115.
Cao, G. and West, M. (1996). Bayesian Analysis of Mixtures. *Biometrics*, 52, 221-227.
Caroll, R.J. and Hall, P. (1988). Optimal Rates of Convergence for Deconvoluting a Density. *Journal of the American Statistical Association*, 83, 1184-1186.
Chen, S.X. (2000). Probability Density Function Estimation Using Gamma Kernels. *Annals of the Institute of Statistical Mathematics*, 52, 471-490.
Dacey, M. F. (1972). A Family of Discrete Probability Distributions Defined by the Generalized Hyper-Geometric Series. *Sankhy $\bar{a}$*, B 34, 243-250.
Dalal, S. R. and Hall, W. J. (1983). Approximating Priors By Mixtures of Natural Conjugate Priors. *Journal of the Royal Statistical Society*, B 45, 278-286.
Dean, C.B., Lawless, J. and Willmot, G.E. (1989). A Mixed Poisson-Inverse Gaussian Regression Model. *Canadian Journal of Statistics*, 17, 171-182.
Devroye, L. (1992). *Non-Uniform Random Variate Generation*. Springer-Verlag. New York.
Diebolt, J. and Robert, C. (1994). Estimation of Finite Mixture Distributions Through Bayesian Sampling. *Journal of the Royal Statistical Society*, B 56, 363-375.
Eberlein, E. and Keller, U. (1995). Hyperbolic Distributions in Finance. *Bernoulli*, 1, 281-299.
Escobar, M. and West, M. (1995). Bayesian Density Estimation and Inference Using Mixtures. *Journal of the American Statistical Association*, 90, 577-588.
Everitt, B.S. (1984). *An Introduction to Latent Variable Models*. Chapman and Hall. New York.
Everitt, B.S. and Hand, D.J. (1981). *Finite Mixtures Distributions*. Chapman and Hall. New York.
Everitt, B.S and Merette, C. (1990). The Clustering of Mixed-Mode Data: A Comparison of possible Approaches. *Journal of Applied Statistics*, 17, 283-297.
Gleser, L. J. (1989). The Gamma Distribution as a Mixture of Exponential Distributions. *American Statistician*, 43, 115-117.
Goldstein, H. (1995). *Multilevel Statistical Models*. (2nd edition) Arnold Publishers. London.
Goutis, K. (1993). Recovering Extra Binomial Variation. *Journal of Statistical Computation and Simulation*, 45, 233-242.
Greenwood, M. and Yule, G. (1920). An Inquiry into the Nature of Frequency Distributions Representative of Multiple Happenings with Particular Reference to the Occurence of Multiple Attacks of Disease or of Repeated Accidents. *Journal of the Royal Statistical Society*, A 83, 255-279.
Gupta, S. and Huang, W.T. (1981). On Mixtures of Distributions: A Survey and Some New Results on Ranking and Selection. *Sankhya*, B 43, 245-290.
Hasselblad, V. (1969) Estimation of Finite Mixtures from the Exponential Family. *Journal of the American Statistical Association*, 64, 1459-1471.
Hebert, J. (1994). Generating Moments of Exponential Scale Mixtures. *Communications in Statistics- Theory and Methods*, 23, 1181-1189.
Hinde, J. and Demetrio, C. G. B. (1998). Overdispersion: Models and Estimation. *Computational Statistics and Data Analysis*, 27, 151-170.
Hosmer, D. (1973). A Comparison of Iterative Maximum Likelihood Estimates of the Parameters of a Mixture of two Normal Distributions Under Three Different Types of Sample. *Biometrics*, 29, 761-770.
Irwin, J. O. (1963). The Place of Mathematics in Medical and Biological Statistics. *Journal of the Royal Statistical Society, A* 126, 1-44.
Irwin, J. O. (1968). The Generalized Waring Distribution Applied to Accident Theory. *Journal of the Royal Statistical Society, A* 131, 205-225.
Irwin, J. O. (1975). The Generalized Waring Distribution. *Journal of the Royal Statistical Society, A* 138, 18-31 (Part I), 204-227 (Part II), 374-384 (Part III).
Jewell, N. (1982). Mixtures of Exponential Distributions. *Annals of Statistics*, 10, 479-484.

Johnson, N.L., Kotz, S.and Kemp, A.W. (1992). *Univariate Discrete Distributions*. (2nd edition) Willey-New York.

Jorgensen, B., Seshadri V. and Whitmore G.A. (1991). On the Mixture of the Inverse Gaussian Distribution With its Complementary Reciprocal. *Scandinavian Journal of Statistics*, 18, 77-89.

Karlin, B. and Lewis, T. (1996). *Empirical Bayes Methods*. Chapman and Hall. New York.

Karlis, D. (1998). Estimation and Hypothesis Testing Problems in Finite Poisson Mixture. *Unpublished Ph.D. Thesis, Dept. of Statistics, Athens University of Economics and Business, ISBN 960-7929-19-5*.

Karlis, D. and Xekalaki, E. (1999). On Testing for the Number of Components in a Mixed Poisson Model. *Annals of the Institute of Statistical Mathematics*, 51, 149-162.

Kendall, M. G. (1961). Natural Law in the Social Sciences. *Journal of the Royal Statistical Society, A* 124, 1-16.

Krishnaji, N. (1970b). Characterization of the Pareto Distribution through a Model of Under-Reported Incomes. *Econometrica*, 38, 251-255.

Laird, N. (1982). Empirical Bayes Estimates Using the Nonparametric Maximum Likelihood Estimate for the Prior. *Journal of Statistical Computation and Simulation*, 15, 211-220.

Lawless, J. (1987). Negative Binomial and Mixed Poisson Regression. *Canadian Journal of Statistics*, 15, 209-225.

Lee, Y. and Nelder, J.A. (1996). Hierarchical Genaralized Linear Models. *Journal of the Royal Statistical Society, B* 58, 619-678.

Leroux, B (1992). Maximum Likelihood Estimation for Hidden Markov Models. *Stochastic Processes*, 49, 127-143.

Lindsay, B. (1983). The Geometry of Mixture Likelihood. A General Theory. *Annals of Statistics*, 11, 86-94.

Lindsay, B. (1995). *Mixture Models: Theory, Geometry and Applications*. Regional Conference Series in Probability and Statistics, Vol. 5, Institute of Mathematical Statistics and American Statistical Association.

Lindsay, B., Clogg, C.C. and Grego, J. (1991). Semiparametric Estimation in the Rasch Model and Related Exponential Response Models, Including a Simple Latent Class for Item Analysis. *Journal of the American Statistical Association*, 86, 96-107.

Liu, M.C. and Taylor, R. (1989). A Consistent Nonparametric Density Estimator for the Deconvolution problem. *Canadian Journal of Statistics*, 17, 427-438.

Mallet, A. (1986). A Maximum Likelihood Estimation Method for Random Coefficient Regression Models. *Biometrika*, 73, 645-656.

Maritz, J. L. and Lwin, (1989). *Empirical Bayes* Methods. (2nd edition) Marcel Dekker Inc. New York.

McLachlan, G. (1992). *Discriminant Analysis and Statistical Pattern Recognition*. Wiley Interscience. New York.

McLachlan, G. and Basford, K. (1989). *Mixture Models: Inference and Application to Clustering*. Marcel Dekker Inc. New York.

McLachlan, G. and Krishnan, T. (1997). *The EM Algorithm and its Extensions*. Wiley. New York.

McLachlan, J.A. and Peel, D. (2001). *Finite Mixture Models*. Wiley. New York.

Mudholkar, G. and Chaubey, Y. P. (1976). On the Distribution of Fisher's Transformation of the Correlation Coefficient. *Communications in Statistics—Computation and Simulation*, 5, 163-172.

Panaretos, J. (1982). An Extension of the Damage Model. *Metrika*, 29, 189-194.

Panaretos, J. (1983). A Generating Model Involving Pascal and Logarithmic Series Distributions. *Communications in Statistics Part A: Theory and Methods*, Vol. A12, No.7, 841-848.

Philippe, A. (1997). Simulation of Right and Left Truncated Gamma Distributions by Mixtures. *Statistics and Computing*, 7, 173-181.

Rachev, St. and Sengupta, A. (1993). Laplace-Weibull Mixtures for Modeling Price Changes. *Management Science*, 39, 1029-1038.

Rao, C.R. (1963). On Discrete Distributions Arising out of Methods of Ascertainments. *Sankhya A*, 25, 311-324.

Redner R., Walker H. (1984). Mixture Densities, Maximum Likelihood and the EM Algorithm. *SIAM Review,* 26, 195-230.

Rudas, T., Clogg, C.C. and Lindsay, B.G. (1994). A New Index of Fit Based on Mixture Methods for the Analysis of Contingency Tables. *Journal of the Royal Statistical Society,* B 56, 623-639.

Scallan, A. J. (1992). Maximum Likelihood Estimation for a Normal/Laplace Mixture Distribution. *The Statistician* 41, 227-231.

Shaked, M. (1980). On Mixtures from Exponential Families. *Journal of the Royal Statistical Society* B 42, 192-198.

Sibuya, M. (1979). Generalised Hypergeometric, Digamma and Trigamma Distributions. *Annals of the Institute of Statistical Mathematics* A 31, 373-390.

Silverman, B.W. (1986). *Density Estimation for Statistics and Data Analysis.* Chapman and Hall. New York.

Simon, H. A. (1955). On a Class of Skew Distribution Functions. *Biometrika,* 42, 425-440.

Simonoff, J.S. (1996). *Smoothing Techniques.* Springer –Verlag. New York.

Symons, M., Grimson, R. and Yuan, Y. (1983). Clustering of Rare Events. *Biometrics* 39, 193-205.

Titterington, D.M. (1990). Some Recent Research in the Analysis of Mixture Distributions. *Statistics* 21, 619-641.

Titterington, D.M. Smith, A.F.M. and Makov, U.E. (1985). *Statistical Analysis of Finite Mixtures Distributions.* Wiley.

Tripathi, R., Gupta, R., Gurland, J. (1994). Estimation of Parameters in the Beta Binomial Model. *Annals of the Institute of Statistical Mathematics,* 46, 317-331.

Wang, M.C. and Van Ryzin, J. (1979). Discrete Density Smoothing Applied to the Empirical Bayes Estimation of a Poisson Mean. *Journal of Statistical Computation and Simulation,* 8, 207-226.

Wang, P., Puterman, M., Cokburn, I. and Le, N. (1996). Mixed Poisson Regression Models with Covariate Dependent Rates. *Biometrics,* 52, 381-400.

Wolfe, J.H. (1970). Pattern Clustering by Multivariate Mixture Analysis. *Multivariate Behavioral Research,* 5, 329-350.

Xekalaki, E. (1983a). A Property of the Yule Distribution and its Applications. *Communications in Statistics, Part A, (Theory and Methods),* A12, 10, 1181-1189.

Xekalaki, E. (1983b). The Univariate Generalised Waring Distribution in Relation to Accident Theory: Proneness, Spells or Contagion? *Biometrics,* 39, 887-895.

Xekalaki, E. (1984a). The Bivariate Generalized Waring Distribution and its Application to Accident Theory. *Journal of the Royal Statistical Society,* A 147, 488-498.

Xekalaki, E. (1984b). Linear Regression and the Yule Distribution. *Journal of Econometrics,* 24 (1), 397-403.

Xekalaki, E. and Panaretos, J. (1983). Identifiability of Compound Poisson Distributions. *Scandinavian Actuarial Journal,* 66, 39-45.

Yule, G. U. (1925). A Mathematical Theory of Evolution Based on the Conclusions of Dr. J.G. Willis. *Philosophical Transactions of the Royal Society,* 213, 21-87.

Received: April 2002, Revised: June 2002

8
NEW PARADIGMS (MODELS) FOR PROBABILITY SAMPLING

Leslie Kish[†]
The University of Michigan

1. Statistics as a New Paradigm

In several sections I discuss new concepts in diverse aspects of sampling, but I feel uncertain whether to call them new paradigms or new models or just new methods. Because of my uncertainty and lack of self-confidence, I ask the readers to choose that term with which they are most comfortable. I prefer to remove the choice of that term from becoming an obstacle to our mutual understanding.

Sampling is a branch of and a tool for statistics, and the field of statistics was founded as a new paradigm in 1810 by Quetelet (Porter, 1987; Stigler, 1986). This was later than the arrival of some sciences: of astronomy, of chemistry, of physics. "At the end of the seventeenth century the philosophical studies of cause and chance...began to move close together.... During the eighteenth and nineteenth centuries the realization grew continually stronger that aggregates of events may obey laws even when individuals do not." (Kendall, 1968). The predictable, meaningful, and useful regularities in the behavior of population aggregates of unpredictable individuals were named "statistics" and were a great discovery.

Thus Quetelet and others computed national (and other) birth rates, death rates, suicide rates, homicide rates, insurance rates, etc. from individual events that are unpredictable. These statistics are basic to fields like demography and sociology. Furthermore, the ideas of statistics were taken later during the nineteenth century also into biology by Frances Galton and Karl Pearson, and into physics by Maxwell, and were developed greatly both in theory and applications.

Statistics and statisticians deal with the effects of chance events on empirical data. The mathematics of chance had been developed centuries earlier for gambling games and for errors of observation in astronomy. Also data have been compiled for commerce, banking, and government. But combining chance with real data needs a new theoretical view, a new paradigm. Thus statistical science and its various branches arrived late in history and in academia, and they are products of the maturity of human development (Kish, 1985).

[†] Leslie Kish passed away on October 7, 2000

The populations of random individuals comprise the most basic concept of statistics. It provides the foundation for distribution theories, inferences, sampling theory, experimental design, etc. And the statistics paradigm differs fundamentally from the deterministic outlook of cause and effect, and of precise relations in the other sciences and mathematics.

2. The Paradigm of Sampling

The Representative Method is the title of an important monograph that almost a century after the birth of statistics and over a century ago now, is generally accepted as the birth of modern sampling (Kiaer, 1895). That term has been used in several landmark papers since then (Jensen, 1926; Neyman, 1934; Kruskal & Mosteller, 1979). The last authors agree that the term "representative" has been used for so many specific methods and with so many meanings that it does not denote any single method. However, as Kiaer used it, and as it is still used generally, it refers to the aims of selecting a sample to represent a population specified in space, in time, and by other definitions, in order to make statistical inferences from the sample to that specified population. Thus a national representative sample demands careful operations for selecting the sample from all elements of the national population, not only from some arbitrary domain such as a "typical" city or province, or from some subset, either defined or undefined.

The scientifically accepted method for survey sampling is probability sampling, which assures known positive probabilities of selection for every element in the frame population. The frame provides the equivalent of listings for each stage of selection. The selection frame for the entire population is needed for mechanical operations of random selection. This is the basis for statistical inferences from the sample statistics to the corresponding population statistics (parameters) (Hansen, Hurwitz, & Madow, 1953, Vol. II). This insistence on inferences based on selections from frame populations is a different paradigm from the unspecified or model based approaches of most statistical analyses.

It took a half century for Kiaer's paper to achieve the wide acceptance of survey sampling. In addition to neglect and passive resistance, there was a great deal of active opposition by national statistical offices who distrusted sampling methods that replaced the complete counts of censuses. Some even preferred the "monograph method," which offered complete counts of a "typical" or "representative" province or district instead of randomly selected national sample (O'Muircheartaigh & Wong, 1981). In addition to political opposition, there were also many opponents among academic disciplines and among academic statisticians. The tide turned in favor of probability sampling with the report of the U.N. Statistical Commission led by Mahanalobis and Yates (U.N., 1950). Five influential textbooks between 1949 and 1954 started a flood of articles with both theory and wide applications.

The strength, the breadth, and the duration of resistance to the concepts and use of probability sampling of frame populations implies that this was a new paradigm that needed a new outlook both by the public and the professionals.

3. Complex Populations

The need for strict probability selection from a population frame for inferences from the sample to a *finite* population is but one distinction of survey sampling. But even more important and difficult problems are caused by the complex distributions of the elements in all the populations. These complexities present a great contrast with the simple model of independence that is assumed, explicitly or implicitly, by almost all statistical theory, all mathematical statistics.

The assumption of independent or uncorrelated observations of variables or elements underlies mathematical statistics and distribution theory. We need not distinguish here between independently and identically distributed (IID) random variables and "exchangeability," and "superpopulations." The simplicity underlying each of those models is necessary for the complexities of the mathematical developments.

Simple models are needed and used for early stages and introductions in all the sciences: for example, perfect circular paths for the planets or $d=gt^2/2$ for freely dropping objects in frictionless situations. But those models fail to meet the complexities of the actual physical worlds. Similarly, independence of elements does not exist in any population whether human, animal, plant, physical, chemical, or biological. The simple independent models may serve well enough for small samples; and the Poisson distribution of deaths by horsekicks in the Prussian Army in 43 years has often served as an example (precious because rare) (Fisher, 1926).

There have also been attempts to construct theoretical populations of IID elements; perhaps the most famous was the classic "collective" of Von Mises (1939); but they do not correspond to actual populations. However, with great effort tables of random numbers have been constructed that have passed all tests. These have been widely used in modern designs of experiments and sample surveys. *Replication* and *randomization* are two of the most basic concepts of modern statistics following the concept of populations.

The simple concept of a population of independent elements does not describe adequately the complex distributions (in space, in time, in classes) of elements. Clustering and stratification are common names for ubiquitous complexities. Furthermore, it appears impossible to form models that would better describe actual populations. The distributions are much too complex and they are also different for every survey variable. These complexities and differences have been investigated and presented now in thousands of computations of "design effects."

Survey sampling needed a new paradigm to deal with the complexities of all kinds of populations for many survey variables and a growing list of survey statistics. This took the form of robust designs of selections and variance formulas that could use a multitude of sample designs, and gave rise to the new discipline of survey sampling. The computation of "design effects" demonstrated the existence, the magnitude, and the variability of effects due to the complexities of distributions not only for means but also for multivariate relations, such as regression coefficients. The long period of disagreements between survey samplers and econometricians testifies to the need for a new paradigm.

4. Combining Population Samples

Samples of national populations always represent subpopulations (domains) which differ in their survey characteristics; sometimes they differ slightly, but at other times greatly. These subclasses in the sample can be distinguished with more or less effort. First, samples of provinces are easily separated when their selections are made separately. Second, subclasses by age, sex, occupation, and education can also be distinguished, and sometimes used for poststratified estimates. Third, however, are those subclasses by social, psychological, and attitudinal characteristics, which may be difficult to distinguish; yet they may be most related to the survey variables. Thus, we recognize that national samples are not simple aggregations of individuals from an IID population, but combinations of subclasses from subpopulations with diverse characteristics. The composition of national populations from diverse domains deserves attention, and it also serves as an example for the two types of combinations that follow. Furthermore, these remarks are pertinent to combinations not only of national samples but also of cities, institutions, establishments, etc.

In recent years two kinds of sample designs have emerged that demand efforts beyond those of simple national samples: (a) periodic samples and (b) multipopulation designs. Each of these has emerged only recently, because they had to await the emergence of three kinds of resources: (1.) effective demand in financial and political resources; (2.) adequate institutional technical resources in national statistical offices; (3.) new methods. In both types of designs we should distinguish the needs of the *survey methods* (definitions, variables, measurements), which must be harmonized, standardized, from *sample designs*, which can be designed freely to fit national (even provincial) situations, provided they are probability designs (Kish, 1994). Both types have been designed first and chiefly for comparisons: periodic comparisons and multinational comparisons, respectively. But new uses have also emerged: "rolling samples" and multinational cumulations, respectively. Each type of cumulation has encountered considerable opposition, and needs a new outlook, a new paradigm.

"Rolling samples" have been used a few times for local situations (Mooney, 1956; Kish, Lovejoy, & Rackow, 1961). Then they have been proposed several times for national annual samples and as a possible replacement for decennial censuses (Kish, 1990). They are now being introduced for national sample censuses first and foremost by the U.S. Census Bureau (Alexander, 1999; Kish, 1990). Recommending this new method, I have usually experienced opposition to the concept of averaging periodic samples: "How can you average samples when these vary between periods?" In my contrary view, the greater the variability the less you should rely on a single period, whether the variation is monotonic, or cyclical, or haphazard. Hence I note two contrasting outlooks, or paradigms. Quite often, the opposition disappears after two days of discussion and cogitation. "For example, annual income is a readily accepted aggregation, and not only for steady incomes but also for occupations with high variations (seasonal or irregular). Averaging weekly samples for annual statistics will prove more easily acceptable than decennial averaging. Nevertheless, many investors in mutual stock funds prefer to rely more on their ten-year or five-year average earnings (despite their obsolescence) than on their up-to-date prior year's earnings (with their risky "random" variations). Most people planning a picnic would also prefer a 50-year average "normal" temperature to last year's exact temperature. There are many similar examples of sophisticated averaging over long periods by the "naïve" public. That public, and policy makers, would also learn fast about rolling samples, given a chance." (Kish, 1998)

Like rolling samples, combining multipopulation samples also encountered opposition: national boundaries denote different historical stages of development, different laws, languages, cultures, customs, religion, and behaviors. How then can you combine them? However, we often find uses and meanings for continental averages; such as European birth and death rates, or South American, or sub-Saharan, or West African rates. Sometimes even world birth, death, and growth rates. Because they have not been discussed, they all usually combined very poorly. But with more adequate theory, they can be combined better (Kish, 1999). But first the need must be recognized with a new paradigm for multinational combinations, followed by developing new and more appropriate methods.

5. Expectation Sampling

Probability sampling assures for each element in the population ($i=1,2,...N$) a known positive probability ($P_i>0$) of selection. The assurance requires some mechanical procedure of chance selection, rather than only assumptions, beliefs, or models about probability distributions. The randomizing procedure requires a practical physical operation that is closely (or exactly) congruent with the probability model (Kish, 1965). Something like this statement appears in most textbooks on survey sampling, and I still believe it all. However, there are two questionable and bothersome objections to this definition and its requirements.

The more important of the two objections concerns the frequent practical situations when we face a choice between probability sampling and expectation sampling. These occur often when the easy, practical selection rate for listing units of $1/F$ yields not only the unique probability $1/F$ for elements, but also some with variable k_i/F for the ith element ($i=1,2,...N$) and with $k_i>0$. Examples of $k_i>1$, usually a small integer, occur with duplicate or replicate lists, dual or multiple frames of selection, second homes for households, mobile populations and nomads, farm operators with multiple lots. Examples of $k_i<1$ are: selecting a single adult from households, selecting single dwellings from buildings. In these examples often the $k_i>1$ or the $k_i<1$ is a small integer and can be easily ascertained, and it is cheaper, more convenient and economical to use weighting than attempting to obtain $1/F$ for all the elements. These problems are described in books and articles.

In most cases, we find it more convenient and less expensive to accept the variable probabilities and to counter them with weighting the expected values $1/k_i$ or k_i, than to operate another stage of selection. Thus, to paraphrase probability sampling: *expectation sampling* assures for each element in the population ($i_1=1,2,...N$) a known positive expectation ($F_i>0$). These procedures are used in practice for descriptive (first order) statistics where the k_i or $1/k_i$ are neither large nor frequent. The treatments for inferential – second order or higher – statistics are more difficult and diverse, and are treated separately in the literature. Note that probability sampling is the special (and often desired) situation when all k_i are 1.

The other objection to the term probability sampling is more theoretical and philosophical and concerns the word "known" in its definition. That word seems to imply belief. Authors from classics like John Venn (1888) and M.G. Kendall (1968) to modern Bayesians like Dennis Lindley – and beyond at both ends – have clearly assigned "probability" to states of belief and "chance" to frequencies generated by objective phenomena and mechanical operations. Thus, our insistence on operations, like random number generators, should imply the term "chance sampling." However, I have not observed its use and it also could lead to a philosophical problem: the proper use of good tables of random numbers implies beliefs in their "known" probabilities. I have spent only a modest amount of time on these problems and agreeable discussions with only a few colleagues, who did agree. I would be grateful for further discussions, suggestions and corrections.

6. Some Related Topics

We called for recognition of new paradigms in six aspects of survey sampling, beginning with statistics itself. Finally, we note here the contrast of sampling to other related methods. Survey methods include the choice and definition of variables, methods of measurements or observations, control of quality (Kish, 1994; Groves, 1989).

Survey sampling has been viewed as a method that competes with censuses (annual or decennial), hence also with registers (Kish, 1990). In some other context, survey sampling competes with or supplements experiments and controlled observations, and clinical trials. These contrasts also need broader comprehensive views (Kish, 1987, Section A.1). However, those discussions would take us well beyond our present limits.

7. References

Alexander, C.H. (1999). A rolling sample survey for yearly and decennial uses. *Bulletin of the International Statistical Institute*, Helsinki, 52nd session.
Fisher, R.A. (1926). *Statistical Methods for Research Workers*. London: Oliver & Boyd.
Groves, R.M. (1989). *Survey Errors and Survey Costs*. New York: John Wiley.
Hansen, M.H., Hurwitz, W.N., and Madow, W.G. (1953). *Sample Survey Methods and Theory*, Vol. I, New York: John Wiley.
Jensen, A. (1926). The representative method in practice, *Bulletin of the International Statistical Institute*, Vol. 22, pt. 1, pp. 359-439.
Kendall, M.G. (1968). Chance, in *Dictionary of History of Ideas*. New York: Chas Scribners.
Kiaer, A.W. (1895). The Representative Method of Statistical Surveys, English translation, 1976. Oslo: Statistik Sentralbyro.
Kish, L. (1965). *Survey Sampling*. New York: John Wiley.
Kish, L. (1981). Using Cumulated Rolling Samples, Washington, Library of Congress.
Kish, L. (1985). Sample Surveys Versus Experiments, Controlled Observations, Censuses, Registers, and Local Studies, *Australian Journal of Statistics*, 27,111-122.
Kish, L. (1987). *Statistical Design for Research*. New York: John Wiley.
Kish, L. (1990). Rolling samples and censuses. *Survey Methodology*, Vol. 16, pp. 63-93.
Kish, L. (1994). Multipopulation survey designs, *International Statistical Review*, Vol. 62, pp. 167-186.
Kish, L. (1998). Space/time variations and rolling samples. *Journal of Official Statistics*, Vol. 14, pp. 31-46.
Kish, L. (1999). Cumulating/combining population surveys. *Survey Methodology*, Vol. 25, pp. 129-138.
Kish, L., Lovejoy, W., and Rackow, P. (1961). A multistage probability sample for continuous traffic surveys, *Proceedings of the American Statistical Association, Section on Social Statistics*, pp. 227-230.
Kruskal, W.H. and Mosteller, F. (1979-80). Representative sampling, I, II, III, and IV, *International Statistical Review*, especially IV: The history of the concept in statistics, 1895-1939.
Mooney, H.W. (1956). *Methodology of Two California Health Surveys*, OS Public Health Monograph 70.
Neyman, J. (1934). On the different aspects of the representative method: the method of stratified sampling and the method of purposive selection. *JRSS*, Vol. 97, pp. 558-625.
O'Muircheartaigh, C. and Wong, S.T. (1981). The impact of sampling theory on survey sampling practice: a review. *Bulletin of International Statistical Institute*, 43rd Session, Vol. 1, pp. 465-493.
Porter, T.M. (1987). *The Rise of Statistical Thinking: 1820-1900*. Princeton, NJ: Princeton University Press.
Stigler, S.M. (1986). *History of Statistics*. Cambridge: Harvard University Press.
United Nations Statistical Office. (1950). *The Preparation of Sample Survey Reports*. New York: UN Series C No. 1; also Revision 2 in 1964.
Venn, J. (1888). *The Logic of Chance*. London: Macmillan.
Von Mises, R. (1939). *Probability, Statistics, and Truth*. London: Wm. Hodge and Co.

(Received: March 2000).
This is the last paper of Leslie Kish. It has also appeared in Survey Methodology, June 2002, No1, 31-34, by the permission of Rhea Kish, Ann Arbor, Michigan.

9
LIMIT DISTRIBUTIONS OF UNCORRELATED BUT DEPENDENT DISTRIBUTIONS ON THE UNIT SQUARE

Samuel Kotz
*Department of Engineering Management
and Systems Engineering
The George Washington University, Washington*

Norman L. Johnson
*Department of Statistics
University of North Carolina at Chapel Hill*

1. Introduction

Recently, bivariate distributions on the unit square $[0,1]^2$ generated by simple modifications of the uniform distribution on $[0,1]^2$ have been studied by the authors of this chapter (Johnson & Kotz, 1998; 1999). These distributions can be used as models for simulating distributions of cell counts in two-way contigency tables (Borkowf et. al., (1997). In our papers we were especially interested in measures of departure from uniformity in distributions with zero correlation between the variables but with some dependence between them. In our 1999 paper we studied a class of distributions that we called *Square Tray* distributions. These distributions are generated by just *two* levels of the probability density function (pdf), but can be extended to multiple-level square tray distributions in the way depicted (for *three* different levels) in Figure 9.1.

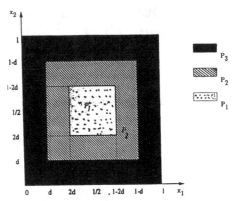

Figure 9.1: Schematic Representation of Multiple Square Tray Distributions.

Let (X_1, X_2) be a bivariate random variable. If its joint density function (pdf) is equal to P_1 in the central square, P_2 in the cross-hatched area, and P_3 in the black region, the distribution of (X_1, X_2) will be called a multiple (three-in this particular case) square tray distribution or *layered square tray distribution* on $[0,1]^2$. (Note that we take $0 < d < 1/4$ as in the case depicted in Figure 9.1).

The pdf of this distribution is:

$$f_X(x_1, x_2) = \begin{cases} P_1 & (x_1, x_2) \in [2d, 1-2d]^2, \\ P_2 & (x_1, x_2) \in [d, d-1]^2 \setminus [2d, 1-2d]^2, \\ P_3 & (x_1, x_2) \in [0,1]^2 \setminus [d, d-1]^2, \\ 0 & \text{otherwise} \end{cases} \quad (1)$$

where $P_j \geq 0, j = 1, 2, 3$. The parameters d, P_1, P_2 and P_3 are connected by the formula:

$$(1-4d)^2 P_1 + 4d(1-3d)P_2 + 4d(1-d)P_3 = 1, \quad (2)$$

or, equivalently,

$$(1-4d)^2 P_{12} + (1-2d)^2 P_{23} + P_3 = 1, \quad (3)$$

where $P_{12} = P_1 - P_2$ and $P_{23} = P_2 - P_3$. The ranges of possible values of $P_j, j = 1, 2, 3$ for a given d are

$$0 \leq P_1 \leq (1-4d)^{-2}, 0 \leq P_2 \leq \frac{1}{4d(1-3d)} \text{ and } 0 \leq P_3 \leq \frac{1}{4d(1-d)}. \quad (4)$$

9. LIMIT DISTRIBUTIONS

2. Pyramid Distributions and their Modifications

In this chapter we first consider the limiting case when the number of different ascending (or descending) square levels increases indefinitely, with the common step height decreasing (or increasing) proportionally. This procedure leads to a continuous distribution on $[0,1]^2$ with pdf constant on the perimeters of concentric squares. A simple subclass of such distributions has pdf's of the form

$$f_{\underset{\sim}{X}}(\underset{\sim}{x}) = a + c \max(|x_1 - \tfrac{1}{2}|, |x_2 - \tfrac{1}{2}|) = a + c\, g(\underset{\sim}{x}) \tag{5}$$

where a and c must satisfy the conditions

$$\int_0^1\!\!\int_0^1 [a + c\, g(\underset{\sim}{x})]\, d\underset{\sim}{x} = a + c\int_0^1\!\!\int_0^1 g(\underset{\sim}{x})\, d\underset{\sim}{x} = 1 \tag{6.1}$$

and

$$a + c\, g(\underset{\sim}{x}) \geq 0 \text{ for all } \underset{\sim}{x} \text{ in } [0,1]^2. \tag{6.2}$$

Since $0 \leq g(\underset{\sim}{x}) \leq \tfrac{1}{2}$, (6.2) means that

$$a \geq 0,\quad a + \tfrac{1}{2} c \geq 0. \tag{6.3}$$

Values of $g(\underset{\sim}{x})$ are exhibited diagrammatically in Figure 9.2. In view of the shape of the graph of $f_{\underset{\sim}{X}}(\underset{\sim}{x})$, we will call distributions with $c < 0$, *pyramid* distributions (on $[0,1]^2$), and those with $c > 0$, *reverse*, or *inverted* pyramid distributions (on $[0,1]^2$).

A natural generalization of these distributions is to take the pdf as

$$f^*_{\underset{\sim}{X}}(\underset{\sim}{x}) = a^* + c^* g^*(\underset{\sim}{x}) \tag{7}$$

where

$$g^*(\underset{\sim}{x}) = \begin{cases} g(\underset{\sim}{x}) & \text{for} \quad g(\underset{\sim}{x}) \geq g_0 \\ g_0 & \text{for} \quad g(\underset{\sim}{x}) \leq g_0 \end{cases} \tag{8}$$

This corresponds to the graph of $f_X^*(x)$ having a flat surface within the square $\max(|x_1 - \frac{1}{2}|, |x_2 - \frac{1}{2}|) = g_0$, i.e., within the square

$$\bigcap_{j=1}^{2} (\frac{1}{2} - g_0 \leq x_j \leq \frac{1}{2} + g_0). \tag{9}$$

See Figure 9.3.

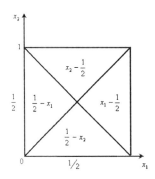
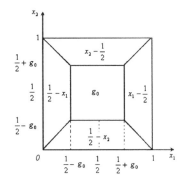

Figure 9.2: Values of $g(x)$ **Figure 9.3:** Values of $g^*(x)$

One might picturesquely describe these distributions as 'Flat-Top' (or 'Mesa') [for $c^* < 0$] and 'Stadium' (for $c^* > 0$) respectively.

We now turn to evaluation of a and c, a* and c*. Since the value of $g(x)$ is constant (=g, say) on the perimeter of squares such as the internal square in Figure 9.3, and the perimeter of this square is

$$4\{\frac{1}{2} + g - (\frac{1}{2} - g)\} = 8g,$$

we have

$$\iint_{0 \leq g(x) \leq g_0} g(x) dx = \int_0^{g_0} g \cdot 8g \, dg = \frac{8}{3} g_0^3. \tag{10}$$

For $g_0 = \frac{1}{2}$

9. LIMIT DISTRIBUTIONS

$$\int_0^1\int_0^1 g(x)\,dx = \frac{1}{3}. \tag{11}$$

Hence for the pyramid and reverse-pyramid distributions ($g_0=0$) we have, from (6.1)

$$a+(1/3)c=1$$

Hence
$$f_X(x) = 1 + \left\{g(x) - \frac{1}{3}\right\}c. \tag{12}$$

From (6.3)
$$-6 \le c \le 3. \tag{13}$$

More generally, when $g_0 > 0$, we have, from (10) and (11),

$$\int_0^1\int_0^1 g^*(x)\,dx = \frac{1}{3} - \frac{8}{3}g_0^3 + (2g_0)^2 g_0 = \frac{1}{3} + \frac{4}{3}g_0^3.$$

Hence
$$a^* + \frac{1}{3}(1+4g_0^3)c^* = 1, \qquad \text{so} \qquad a^* = 1 - \frac{1}{3}(1+4g_0^3)c^*, \tag{14}$$

and
$$f_X^*(x) = 1 + \left\{g^*(x) - \frac{1}{3}(1+4g_0^3)\right\}c^*. \tag{15}$$

Since $g_0 \le g^*(x) \le \frac{1}{2}$,

$$1 + \left\{g_0 - \frac{1}{3}(1+4g_0^3)\right\}c^* \ge 0$$

and
$$1 + \left(\frac{1}{6} - \frac{4}{3}g_0^3\right)c^* \ge 0,$$

so that
$$-\left(\frac{1}{6} - \frac{4}{3}g_0^3\right)^{-1} \le c^* \le \left\{\frac{1}{3}(1+4g_0^3) - g_0\right\}^{-1}. \tag{16}$$

(Note that $\frac{1}{6} - \frac{4}{3}g_0^3 \ge 0$ and $\frac{1}{3}(1+4g_0^3) - g_0 \ge 0$ for $0 \le g_0 \le \frac{1}{2}$).

3. Structural Properties

We shall now focus our attention on square pyramid and inverted pyramid distributions ($g_0=0$). First, we evaluate the pdf of X_1. We have, for $0<x_1<\frac{1}{2}$,

$$f_{X_1}(x_1) = \int_0^1 f_X(x_1,x_2)\, dx_2 \qquad (16.1)$$

$$= 1 + \left\{ \int_0^{x_1} \left(\frac{1}{2} - t_2\right) dt_2 + \int_{x_1}^{1-x_1} \left(\frac{1}{2} - x_1\right) dt_2 + \int_{1-x_1}^1 \left(t_2 - \frac{1}{2}\right) dt_2 - \frac{1}{3} \right\} c$$

$$= 1 + \left\{ 2\int_0^{x_1} \left(\frac{1}{2} - t_2\right) dt_2 + \left(\frac{1}{2} - x_1\right)(1 - 2x_1) - \frac{1}{3} \right\} c$$

$$= 1 + \left\{ \frac{1}{4} - \left(\frac{1}{2} - x_1\right)^2 + \frac{1}{2}(1 - 2x_1)^2 - \frac{1}{3} \right\} c$$

$$= 1 + \left\{ \frac{1}{4}(1 - 2x_1)^2 - \frac{1}{12} \right\} c = 1 + \left\{ \frac{1}{6} - x_1(1 - x_1) \right\} c$$

This formula is also valid for $\frac{1}{2} \le x_1 \le 1$. Similarly

$$f_{X_2}(x_2) = 1 + \left\{ \frac{1}{6} - x_2(1 - x_2) \right\} c \qquad (0 < x_2 < 1). \qquad (16.2)$$

From equations (16)

$$E[X_j] = \frac{1}{2} \quad \text{and} \quad \text{var}(X_j) = \frac{1}{12} + \frac{1}{180} c \qquad (j=1,2). \qquad (17)$$

The cumulative distribution function (cdf) of X_j is

$$F_{X_j}(x_j) = \int_0^{x_j} \left[1 + \left\{ \frac{1}{6} - t(1-t) \right\} c \right] dt = x_j + \left(\frac{1}{6} x_j - \frac{1}{2} x_j^2 + \frac{1}{3} x_j^3 \right) c$$

$$= x_j \left\{ 1 + \frac{1}{6}(1 - x_j)(1 - 2x_j)\, c \right\}. \qquad (18)$$

9. LIMIT DISTRIBUTIONS

To evaluate the joint cdf, $F_X(\underset{\sim}{x}) = \Pr\left[\bigcap_{j=1}^{2}(X_j \leq x_j)\right]$, we first take the case $0 \leq x_1 \leq x_2 \leq \frac{1}{2}$, as is represented in Figure 9.4 (p.110). The region $\bigcap_{j=1}^{2}(0 \leq t_j \leq x_j)$ is split into subregions

R_1: $(0 \leq t_1 \leq x_1) \cap (x_1 \leq t_2 \leq x_2)$,

R_2: $0 \leq t_1 \leq t_2 \leq x_1$,

and R_2': $0 \leq t_2 \leq t_1 \leq x_1$.

Then
$$F_X(\underset{\sim}{x}) = \iint_{R_1} + \iint_{R_2} + \iint_{R_2'} f_X(t_1, t_2) d\underset{\sim}{t}$$

$$= x_1 x_2 \left(1 - \frac{1}{3}c\right) + \left\{\iint_{R_1} + 2\iint_{R_2} g(t_1, t_2) d\underset{\sim}{t}\right\}c$$

(since $f_X(\underset{\sim}{t}) = 1 + \left\{g(\underset{\sim}{t}) - \frac{1}{3}c\right\}$ and $\iint_{R_2} g(\underset{\sim}{t}) d\underset{\sim}{t} = \iint_{R_2'} g(\underset{\sim}{t}) d\underset{\sim}{t}$).

Now $\iint_{R_1} g(\underset{\sim}{t}) d\underset{\sim}{t} = \int_0^{x_1} \int_{x_1}^{x_2} \left(\frac{1}{2} - t_1\right) dt_2 dt_1 = \frac{1}{2}(x_2 - x_1)x_1(1 - x_1)$

and

$\iint_{R_2} g(\underset{\sim}{t}) d\underset{\sim}{t} = \int_0^{x_1} \int_{t_1}^{x_1} \left(\frac{1}{2} - t_1\right) dt_1 = \frac{1}{2}x_1^2 - \frac{1}{4}x_1^2 - \frac{1}{2}x_1^3 + \frac{1}{3}x_1^3 = \frac{1}{4}x_1^2 - \frac{1}{6}x_1^3$.

Hence
$$F_X(\underset{\sim}{x}) = x_1 x_2 (1 - \frac{1}{3}c) + \{\frac{1}{2}(x_2 - x_1)x_1(1 - x_1) + \frac{1}{2}x_1^2 - \frac{1}{2}x_1^3\}c$$

$$= x_1 x_2 + \frac{1}{6}\{(1 - 3x_1)x_2 + x_1^2\}c. \qquad (0 \leq x_1 \leq x_2 \leq \frac{1}{2}). \qquad (19)$$

For $0 \leq x_2 \leq x_1 \leq \frac{1}{2}$, interchange subscripts 1 and 2.

For $x_1 \leq \frac{1}{2} \leq x_2$, $F_X(x_1, x_2) = 1 - F_X(x_1, 1 - x_2)$.

For $\frac{1}{2} \leq x_2 \leq x_1 \leq 1$, $F_X(x_1, x_2) = 1 - F_X(x_1) - F_X(x_2) + F_X(1 - x_1, 1 - x_2)$

and so on.

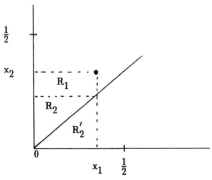

Figure 9.4: Evaluation of $F_X(x)$ for $0 \leq x_1 \leq x_2 \leq \frac{1}{2}$

From symmetry

$$E[X_1 | X_2 = x_2] = E[X_2 | X_1 = x_1] = \frac{1}{2}, \text{ for all } x \text{ in } [0,1]^2.$$

Hence the correlation between X_1 and X_2 is zero. However, the two variables are not mutually independent. (Evidently $f_{X_1}(x_1) \neq f_{X_1|X_2}(x_1 | x_2)$).

It might be of interest to note that $f_{X_1|X_2}(x_1 | x_2) = \dfrac{1 + \left\{ g(x) - \frac{1}{3} \right\} c}{1 + \left\{ \frac{1}{6} - x_2(1 - x_2) \right\} c}$

is constant (with respect to x_1) over all values of x_1 for which $g(x) = \left| x_2 - \frac{1}{2} \right|$

i.e., for which $\left| x_1 - \frac{1}{2} \right| \leq \left| x_2 - \frac{1}{2} \right|$.

(See Figure 9.5).

9. LIMIT DISTRIBUTIONS

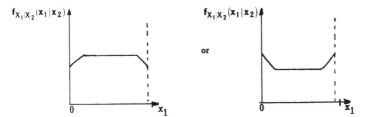

Figure 9.5: Schematic representation of conditional densities

Further to (15) and (16) we have (cf.(7)):

$$\left(f_{X_1}^*(x_1) - a^*\right)/c^* = \begin{bmatrix} 2\int_0^{x_1}\left(\frac{1}{2} - t_1\right)dt_2 + \int_{x_1}^{1-x_1}\left(\frac{1}{2} - x_1\right)dt_2 \\ 2\int_0^{\frac{1}{2}-g_0}\left(\frac{1}{2} - t_1\right)dt_2 + \{g_0 \times (2g_0)\} \end{bmatrix}$$

$$= \begin{bmatrix} \frac{1}{4} + \frac{1}{4}(1-2x_1)^2 & \text{for} & \left|\frac{1}{2} - x_1\right| > g_0 \\ \frac{1}{4} + g_0^2 & \text{for} & \frac{1}{2} - g_0 \leq x_1 \leq \frac{1}{2} + g_0 \end{bmatrix}$$

leading to

$$f_{X_j}^*(x_j) = \begin{bmatrix} 1 + \left\{\frac{1}{6} - \frac{4}{3}g_0^3 - x_j(1-x_j)\right\}c^* & \text{for} & \left|\frac{1}{2} - x_1\right| > g_0 \\ 1 + \left(g_0^2 - \frac{1}{12} - \frac{4}{3}g_0^3\right)c^* & \text{for} & \frac{1}{2} - g_0 \leq x_1 \leq \frac{1}{2} + g_0 \end{bmatrix} \quad (20)$$

Note that $f_{X_j}^*(x_j)$ is constant for $\frac{1}{2} - g_0 \leq x_1 \leq \frac{1}{2} + g_0$.

For $\bigcup_{j=1}^{2}\left(0 \leq x_j \leq \frac{1}{2} - g_0\right)$

$$F_X^*(x) = a^* x_1 x_2 + \frac{F_X(x) - ax_1x_2}{c} c^*. \quad (21.1)$$

For $\bigcap_{j=1}^{2}(\frac{1}{2} - g_0 \leq x_j \leq \frac{1}{2} + g_0)$

$$F_X^*\left(\underset{\sim}{x}\right) = g_0 c^* \prod_{j=1}^{2}\left(x_j - \frac{1}{2} + g_0\right) + a^* x_1 x_2$$

$$+ c^{-1}\left[F_X\left(\frac{1}{2} - g_0, x_2\right) - a\left(\frac{1}{2} - g_0\right)x_1 + F_X\left(x_1, \frac{1}{2} - g_0\right) - a\left(\frac{1}{2} - g_0\right)x_2\right.$$

$$\left. - F_X\left(\frac{1}{2} - g_0, \frac{1}{2} - g_0\right) + a\left(\frac{1}{2} - g_0\right)^2\right]c^*$$

$$= a^* x_1 x_2 + g_0 c^* \prod_{j=1}^{2}\left(x_j - \frac{1}{2} + g_0\right)$$

$$+ c^{-1}\left[F_X\left(\frac{1}{2} - g_0, x_2\right) + F_X\left(x_1, \frac{1}{2} - g_0\right) - F_X\left(\frac{1}{2} - g_0, \frac{1}{2} - g_0\right)\right.$$

$$\left. - a\left(\frac{1}{2} - g_0\right)\left(x_1 + x_2 - \frac{1}{2} + g_0\right)\right]c^*. \quad (21.2)$$

4. Measures of Departure from Uniformity

We also have the measures of departure from uniformity

$$\sigma = 12 \int_0^1 \int_0^1 |F_X(\underset{\sim}{x}) - x_1 x_2| \, d\underset{\sim}{x} \quad (22.1)$$

and

$$\delta = \int_0^1 \int_0^1 |F_X(\underset{\sim}{x}) - 1| \, d\underset{\sim}{x}. \quad (22.2)$$

Details of the calculations are given below. Measure (22.1) is related to the measure of *dependence* introduced by Schweizer and Wolff (1976; 1981). It would be the same as that measure if both marginal distributions were uniform. Measure (22.2) is similarly related to the measure introduced by Feuerverger (1993).

a) Calculation of $\int_0^1 \int_0^1 |F_X(\underset{\sim}{x}) - x_1 x_2| \, d\underset{\sim}{x}$ for pyramid-type square distributions.

We have

$$f_X(\underset{\sim}{x}) = 1 + \left\{\max\left(|x_1 - \frac{1}{2}|, |x_2 - \frac{1}{2}|\right) - \frac{1}{3}\right\}c, \quad \left(\underset{\sim}{x} \in [0,1]^2, -6 \le c \le 3\right).$$

9. LIMIT DISTRIBUTIONS

$$f_{x_j}(x_j) = 1 + \left\{\frac{1}{6} - x_j(1-x_j)\right\}c$$

and

$$F_{X_j}(x_j) = x_j\left\{1 + \frac{1}{6}(1-x_j)(1-2x_j)c\right\}. \qquad (j=1,2;\ a < x_j \le 1).$$

In the octant $0 \le x_1 \le x_2 \le \frac{1}{2}$, $F(x) = x_1 x_2 + \frac{1}{6}x_1\{(1-3x_1)x_2 + x_1^2\}c$. The contributions to

$$\int_0^1\int_0^1 |F_X(x) - x_1 x_2| \, dx$$

are the same for each of the 8 octants:

$0 \le x_1 \le x_2 \le \frac{1}{2};\ 0 \le x_1 \le \frac{1}{2} \le x_2 \le 1 - x_1$, etc. (See Figure 9.6).

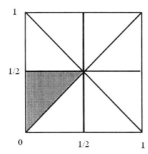

Figure 9.6: Subdivision of the unit square into 8 octants

We shall concentrate on the octant $0 \le x_1 \le x_2 \le \frac{1}{2}$. The integral over $[0,1]^2$ is 8 times the integral over this (or any other) octant. We note that in this octant

$$h(x) \equiv (1 - 3x_1) x_2 + x_1^2 \ge 0.$$

Indeed $\dfrac{\partial h(x)}{\partial x_2} = 1 - 3x_1$ has for the same sign for all x_2. Since

$h(x_1, x_1) = 1 - 2x_1^2 \ge 0$ for $0 \le x_1 \le \frac{1}{2}$ and

$h\left(x_1, \frac{1}{2}\right) = \frac{1}{2} - \frac{3}{2}x_1 + x_1^2 = \frac{1}{2}(1-x_1)(1-2x_1) \ge 0$ for $0 \le x_1 \le \frac{1}{2}$.

Direct calculations show that

$$\sigma = 12 \int_0^1 \int_0^1 |F_X(x) - x_1 x_2| \, dx = \frac{96|c|}{6} \int_0^{\frac{1}{2}} \int_0^{\frac{1}{2}} t_1 \{(1-3t_1)t_2 + t_1^2\} dt_2 dt_1 = \frac{|c|}{120}.$$

Comparing with the situation for the "uniform" Farlie-Gumbel-Morgenstern distribution (see e.g., Johnson & Kotz, 1975) where

$$F_X(x) = x_1 x_2 [1 + \alpha\{(1-x_1)(1-x_2)\}] \qquad x \in [0,1]^2, \; -1 \le \alpha \le 1$$

for which—using our notation— $\sigma = \frac{1}{3}|\alpha|$, we note that the distribution at hand has remarkably weak "dependence" between the components, as measured by deviation from bivariate uniformity on the unit square. Note also that for square tray distributions, the value of σ in the vicinity of $d=0.25$ is 0.3 (Johnson & Kotz, 1999).

b) Evaluation of $\int_0^1 \int_0^1 |f_X(x) - 1| \, dx$ yields:

$$\int_0^1 \int_0^1 |f_X(x) - 1| \, dx = |c| \, 8 \int_0^{\frac{1}{2}} \int_{t_1}^{\frac{1}{2}} |\frac{1}{6} - t_1| \, dt_2 dt_1.$$

[Since $f_X(x) = 1 + \left(\frac{1}{6} - x_1\right)c$ for $0 \le x_1 \le x_2 \le \frac{1}{2}$].

Now $\int_0^{\frac{1}{6}} \int_{t_1}^{\frac{1}{2}} \left(\frac{1}{6} - t_1\right) dt_2 \, dt_1 + \int_{\frac{1}{6}}^{\frac{1}{2}} \int_{t_1}^{\frac{1}{2}} \left(t_1 - \frac{1}{6}\right) dt_2 \, dt_1 = \frac{1}{81}$

and $\delta = \int_0^1 \int_0^1 |f_X(x) - 1| \, dx = \frac{8}{81}|c|. \qquad (-6 \le c \le 3).$

This value is also quite low compared with those for the square tray distributions where $\int_0^1 \int_0^1 |f_X(x) - 1| \, dx = 8d(1-d)(P_2 - 1)$ which for $P_2=1.5$ and $d=1/4$ gives $3/4$ (Johnson & Kotz, 1999). Even $d=1/10$ results in the value of 0.36 for the square tray distribution, more than 3 times larger than the value of the same measure for the limiting distribution with $c=1$.

9. LIMIT DISTRIBUTIONS

We note that the value of δ for square 'Flat-Top' ('Mesa') and 'Stadium' distributions can be obtained by recalling that

$$c^{*-1}\left\{f_X^*(\underset{\sim}{x})-1\right\} = \begin{cases} \dfrac{1}{6}-x_1-\dfrac{4}{3}g_0^3 & \text{for } x_1 \leq \dfrac{1}{6}-\dfrac{4}{3}g_0^2 \\ g_0-\dfrac{1}{3}-\dfrac{4}{3}g_0^2 & \text{for } \dfrac{1}{6}-\dfrac{4}{3}g_0^2 \leq x_1 \leq \dfrac{1}{2}. \end{cases}$$

Evidently $f_X(\underset{\sim}{x})-1>0$ if $x_1 < \dfrac{1}{6}-\dfrac{4}{3}g_0^2$ or $f_X(\underset{\sim}{x})-1<0$ if $x_1 > \dfrac{1}{6}-\dfrac{4}{3}g_0^2$.

Hence

$$c^{*-1}\int_0^{\frac{1}{2}}\int_0^{\frac{1}{2}}|f_X^*(\underset{\sim}{t})-1|\,dt = \int_0^{\frac{1}{6}-\frac{4}{3}g_0^2}(\frac{1}{2}-t_1)(\frac{1}{6}-t_1-\frac{4}{3}g_0^3)dt_1$$

$$-\int_{\frac{1}{6}-\frac{4}{3}g_0^2}^{\frac{1}{2}-g_0}(\frac{1}{2}-t_1)(g_0-\frac{1}{3}-\frac{4}{3}g_0^2)\,dt_1$$

$$-\int_{\frac{1}{2}-g_0}^{\frac{1}{2}}(\frac{1}{2}-t_1)(g_0-\frac{1}{3}-\frac{4}{3}g_0^2)dt_1$$

and

$$\delta = \int_0^1\int_0^1|f_X^*(\underset{\sim}{x})-1|\,dx = 8\left\{\frac{1}{81}(1+4g_0^3)^3 - \frac{1}{3}g_0^3\right\}|c^*|.$$

Note that δ is a linear function of g_0^3.

The following table shows that $\delta/|c^*|$ is a decreasing function of g_0. For $g_0=0$ (the pyramid distribution) we have $\delta/|c^*|=\dfrac{8}{81}=0.09877$ and for $g_0=\dfrac{1}{2}$ (the uniform case) the value is 0 corresponding to independence.

g_0	$\delta/|c^*|$
0	0.09877
0.1	0.09729
0.2	0.08722
0.3	0.06235
0.4	0.02503
0.5	0.00000

Indeed: set $h(g_0^3) = \dfrac{\delta}{8|c^*|}$ and denote $g_0^3 = y$, then

$$h(y) = \frac{1}{81}(1+4y)^3 - \frac{1}{3}y$$

and

$$h'(y) = \frac{12}{81}(1+4y)^2 - \frac{1}{3}.$$

$h(y)$ is a decreasing function of y for $0 \le y \le \left(\dfrac{1}{2}\right)^3 = \dfrac{1}{8}$ (i.e. $0 \le g_0 \le \dfrac{1}{2}$) and $h\left(\dfrac{1}{8}\right) = 0$; $h'(y) < 0$ for $0 \le y < \dfrac{1}{8}$.

5. Concluding Remarks

The limiting distribution of layered tray distributions on $[0,1]^2$ is found to be a distribution with a rather weak dependence among components (compared with the initial tray distributions). Relatively simple structure of this distribution with appealing and flexible marginals will no doubt be useful in applications in statistics and applied probability, in which bivariate models become prominent. Nevertheless the limiting process seems to be not sufficiently powerful to yield a distribution with (limiting) independent components!

Acknowledgments

Most sincere thanks are due to Mrs. Teresita R. Abacan for her skillful typing.

6. References

Borkowf, C. B., M. H. Gail, R. J. Carroll and R. D. Gill (1997). Analyzing bivariate continuous data grouped into categories defined by empirical quantiles of marginal distributions, *Biometrics*, 53:1054-1069.

Feuerverger, A. (1993). A consistent test for bivariate dependence, *International Statistical Review*, 61: 419-433.

Johnson, N. L. and S. Kotz (1975). On some generalized Farlie-Gumbel-Morgenstern distributions. *Communication in Statistics*, 4 :411-427.

Johnson, N. L. and S. Kotz (1998, 1999). *Square Tray Distribution*. TR-4-10-98. The George Washington University, Operations Research Department, Washington, DC 20052. (Appeared in *Statistics and Probability Letters*, 42: 157-165).

Schweizer, B. and E. F. Wolff (1976). Sur une mesure de dependence pour les variables aleatoires, *Comptes Rendus de l' Academie des Sciences, Paris*, 233A: 659-661.

Schweizer, B. and E. F. Wolff (1981). On nonparametric measures of dependence for random variables, *Annals of Statistics*, 9:879-885.

Received: July 1999

10
LATENT VARIABLE MODELS WITH COVARIATES

Irini Moustaki
Department of Statistics
Athens University of Economics and Busines, Greece

1. Introduction

Survey data contains variables which are measured on a categorical scale (nominal or ordinal), or a metric scale (discrete or continuous) or combinations of the above. A latent variable model fitted to such data must take into account the scale of each variable.

In the literature there are two approaches for fitting latent variable models. First, there is the underlying variable approach that treats all observed variables as continuous by assuming that underlying each categorical observed variable there is a continuous unobserved variable. Within this approach, contributions have been made by Jöreskog and Sörbom (1993), Muthen (1984), Lee, Poon and Bentler (1992), and Arminger and Kusters (1988). Secondly, there is the response function approach that starts by defining for each individual in the sample the probability of responding positively to a variable given the individual's position on the latent factor space. Within this approach authors such as Lawley and Maxwell (1971), Bock and Aitkin (1981), Bartholomew and Knott (1999), and many other psychometricians have looked at models for either categorical or metric variables. The approach used in this chapter is an extension of the Moustaki and Knott (2000) paper on categorical and metric variables and Moustaki (1996) for mixed (binary and metric) manifest variables and it is fundamentally different from the underlying variable approach as will be shown.

Sammel, Ryan, and Legler (1997) discuss a latent trait model for mixed outcomes with covariate effects. Results in their paper are restricted to binary and normal manifest outcomes. In our paper, we model the effect of latent variables and observed covariates on manifest variables from any distribution in the exponential family. In addition to binary and normal outcomes, our results cover polytomous, Poisson, and gamma distributed variables.

Generalized linear models (GLIM) were introduced by Nelder and Wedderburn (1972) and a systematic discussion of them can be found in McCullagh and Nelder (1989). GLIM includes as special cases linear regression models with Normal, Poisson, or Binomial errors and log-linear models. In all these models the explanatory variables are observed variables. In our results some of the explanatory variables are latent (unobserved) variables.

We put the latent variable model with mixed manifest variables in a general framework. A program called GENLAT (Moustaki, 2002a) has been written in FORTRAN 77 to fit all the models proposed in the paper.

The chapter is organized as follows: Section 2 discusses the theoretical framework of generalized linear models; Section 3 discusses the estimation method used for the generalized latent variable model with covariates; Section 4 illustrates the methodology using two real examples, and finally, Section 5 outlines the results of this work.

2. Generalized linear models

A generalized linear model consists of three components:

a. The random component in which each of the p random response variables, $(y_1, \ldots, y_p)$ has a conditional distribution from the exponential family, (such as Bernoulli, Poisson, Multinomial, Normal, Gamma).

b. The systematic component in which covariates, here both explanatory observed variables $\mathbf{x}' = (x_1, x_2, \ldots, x_r)$ and latent variables $\mathbf{z}' = (z_1, z_2, \ldots, z_q)$ produce a linear predictor η_i corresponding to each y_i:

$$\eta_i(z, x) = \alpha_{i0} + \sum_{j=1}^{q} \alpha_{ij} z_j + \sum_{j=1}^{r} \beta_{ij} x_j, \quad i=1,\ldots,p \quad (1)$$

c. The links between the systematic component and the conditional means of the random component distributions: $\eta_i = u_i(\mu_i)$ where $\mu_i = E(y_i \mid z, x)$ and $u_i(.)$ is called the link function which can be any monotonic differentiable function and may be different for different manifest variables y_i, $i = 1,\ldots,p$.

We shall, in fact, assume that $(y_1, y_2, \ldots, y_p)$ denotes a vector of p manifest variables where each variable has a conditional distribution in the exponential family taking the form:

$$f_i(y_i \mid z, x) = \exp\left\{ \frac{y_i \theta_i(z, x) - b_i(\theta_i(z, x))}{\varphi_i} + d_i(y_i, \varphi_i) \right\}, i=1,\ldots,p \quad (2)$$

where $b_i(\theta_i(z,x))$ and $d_i(y_i, \phi_i)$ are specific functions taking a different form depending on the distribution of the response variable y_i. All the distributions discussed in this chapter have canonical link functions with:

$$\theta_i(z,x) = \eta_i(z,x)$$

and ϕ_i is a scale parameter.

Table 10.1 gives for several different types of responses the form of the specific functions and the link function of the generalized model. For simplicity of exposition we illustrate the models with a single latent variable, z. The results are easily extended to any number of latent variables (Moustaki & Knott, 2000).

10. LATENT VARIABLE MODELS

Table 10.1: Components of the generalized latent variable model with covariates

	Link	$b_i(\theta_i(z,x))$	$d_i(y_i, \phi_i)$
Binary	Logit	$\log(1 + e^{\theta_i(z,x)})$	0
Polytomous	Logit	$\log(\Sigma_{s=1}^{c_i} e^{\theta_{i(s)}(z,x)})$	0
Poisson	Log	$\exp(\theta_i(z,x))$	0
Normal	Identity	$[\theta_i(z,x)]^2/2$	$-\dfrac{1}{2}\left[\dfrac{y_i^2}{\phi_i} + \log(2\pi\phi_i)\right]$
Gamma	Inverse	$-\log(-\theta_i(z,x))$	$v_i \log(v_i y_i) - \log y_i - \log \Gamma(v_i)$

More specifically:

Binary items

In case of binary responses, y_i takes values 0 and 1. Suppose that the manifest binary variable has a Bernoulli distribution with expected value $\pi_i(z, x)$. The link function is defined to be the logit, i.e.,:

$$\text{logit}\,\pi_i(z,\mathbf{x}) = \alpha_{i0} + \alpha_{i1}z + \sum_{j=1}^{r} \beta_{ij}x_j, \; i=1,\ldots,p$$

where

$$\pi_i(z,\mathbf{x}) = P(y_i = 1 \mid z, \mathbf{x}) = e^{\theta_i(z,x)}/(1 + e^{\theta_i(z,x)})$$

and the conditional distribution of $y_i \mid z$ is given by:

$$g_i(y_i \mid z, \mathbf{x}) = \pi_i(z,\mathbf{x})^{y_i}(1 - \pi_i(z,\mathbf{x}))^{1-y_i}. \qquad (3)$$

Polytomous items

In the polytomous case, the indicator variable y_i is replaced by a vector-valued indicator function with its s'th element defined as:

$$y_{i(s)} = \begin{cases} 1, & \text{if the response falls in category } s, \text{ for } s = 1,\ldots,c_i \\ 0, & \text{otherwise} \end{cases}$$

where c_i denotes the number of categories of variable i and $\Sigma_{s=1}^{c_i} y_{i(s)} = 1$. The response pattern of an individual is written as $\mathbf{y'} = (\mathbf{y'}_1, \mathbf{y'}_2, \ldots, \mathbf{y'}_p)$ of dimension $\Sigma_i c_i$. This chapter discusses only variables measured on a nominal scale for the ordinal case see Moustaki (2002b).

The single response function of the binary case is now replaced by a set of functions $\pi_{i(s)}(z,x)$ ($s = 1, \ldots, c_i$) where $\sum_{s=1}^{c_i}\pi_{i(s)}(z,x)=1$.

In the binary case, both y_i and θ_i are scalars where in the polytomous case there are vectors. The first category of the polytomous variable is arbitrarily selected to be the reference category. The vector $\theta_i(z, x)$ is written as:

$$\theta_i'(z,x) = \left\{0, \ln\frac{\pi_{i(2)}(z,x)}{\pi_{i(1)}(z,x)}, \ldots, \ln\frac{\pi_{i(c_i)}(z,x)}{\pi_{i(1)}(z,x)}\right\}, \quad i=1,\ldots,p$$

The canonical parameter $\theta_i(z, x)$ remains a linear function of the latent variable:

$$\theta_i(z,x) = \alpha_i + \alpha_{i1}z + \sum_{j=1}^{r}\beta_{ij}x_j$$

where $\quad \alpha_{il}' = (\alpha_{il(1)} = 0, \alpha_{il(2)}, \ldots, \alpha_{il(c_i)}), l = 0,1$

and $\quad \beta_{ij}' = (\beta_{ij(1)} = 0, \beta_{ij(2)}, \ldots, \beta_{ij(c_i)}), j = 1, \ldots, r.$

As $\pi_{i(s)}(z)$ is over-parameterized, we fix the parameters of the first category to zero, $\alpha_{i0(1)} = \alpha_{i1(1)} = \beta_{ij(1)} = 0$ for all j. The conditional distribution of y_i given z is taken to be the multinomial distribution:

$$g_i(y_i \mid z) = \prod_{s=1}^{c_i}(\pi_{i(s)}(z,x))^{y_{i(s)}}, \qquad (4)$$

Normal items

In the normal case $y_i|z,x$ has a normal distribution with mean $\alpha_{i0} + \alpha_{i1}z + \sum_{j=1}^{r}\beta_{ij}x_j$ and variance ϕ_i.

Gamma items

In the gamma case the shape parameter is $v_i = \dfrac{1}{\phi_i}$ and the dispersion parameter is $\dfrac{\gamma_i}{v_i} = \gamma_i\phi_i$.

3. Estimation

We estimate the parameters by maximum likelihood based on the joint distribution of the manifest variables. In this formulation of the model we allow the manifest variables to take any form from the exponential family.

10. LATENT VARIABLE MODELS

Under the assumption of local independence (responses to the p manifest items are independent conditional on the latent variables and the set of explanatory variables x) the joint distribution of the manifest variables is:

$$f(y \mid x) = \int_{-\infty}^{+\infty} g(y \mid z, x) h(z) dz = \int_{-\infty}^{+\infty} \left[\prod_{i=1}^{p} g_i(y_i \mid z, x) \right] h(z) dz.$$

$g_i(y_i \mid z, x)$ can be any distribution from the exponential family. The latent variable z is taken to be standard normal and $h(z)$ denotes the standard normal density.

For a random sample of size n the log-likelihood is written as:

$$L = \sum_{m=1}^{n} \log f(y_m \mid x_m)$$

$$= \sum_{m=1}^{n} \log \int_{-\infty}^{+\infty} g(y_m \mid z, x_m) h(z) dz$$

$$= \sum_{m=1}^{n} \log \int_{-\infty}^{+\infty} \left[\prod_{i=1}^{p} \exp\left\{ \frac{y_{im} \theta_i(z, x_m)}{\varphi_i} - \frac{b_i(\theta_i(z, x_m))}{\varphi_i} + d_i(\varphi_i, y_{im}) \right\} \right] h(z) dz. \quad (5)$$

The unknown parameters are in $\theta_i(z,x_m)$, $b_i(\theta_i(z,x_m))$ and φ_i. We differentiate the log-likelihood given in equation (5) with respect to the model parameters, α_{i0}, α_{i1}, β_{ij} and the scale parameter φ_i where $i=1,...,p$ and $j=1,...,r$.

We denote with $\boldsymbol{a}'_i = (\alpha_{i0}, \alpha_{i1}, \beta_{i1},...\beta_{ir})$ the vector with all the unknown parameters for item i.

Finding partial derivatives, we have

$$\frac{\partial L}{\partial a_i} = \sum_{m=1}^{n} \frac{1}{f(y_m \mid x_m)} \frac{\partial f(y_m \mid x_m)}{\partial a_i}$$

$$= \sum_{m=1}^{n} \frac{1}{f(y_m \mid x_m)} \int_{-\infty}^{+\infty} g(y_m \mid z, x_m) h(z) \frac{\partial}{\partial a_i} \left[\frac{y_{im} \theta_i(z, x_m)}{\varphi_i} - \frac{b_i(\theta_i(z, x_m))}{\varphi_i} \right] dz. \quad (6)$$

The integral in equation (6) can be approximated by Gauss-Hermite quadrature with weights $h(z_t)$ at abscissae z_t. By interchanging the summation we get:

$$\frac{\partial L}{\partial a_i} = \sum_{t=1}^{k} \frac{h(z_t)}{\varphi_i} \left[\sum_{m=1}^{n} y_{im} \frac{g(y_m \mid z_t, x_m)}{f(y_m \mid x_m)} \frac{\partial \theta_i(z_t, x_h)}{\partial a_i} - \sum_{m=1}^{n} \frac{g(y_m \mid z_t, x_m)}{f(y_m \mid x_m)} \frac{\partial b_i \theta_i(z_t, x_m)}{\partial a_i} \right] \quad (7)$$

For example, in the binary case the partial derivatives with respect to the model parameters after they are put equal to zero are:

$$\frac{\partial L}{\partial a_{il}} = \sum_{t=1}^{k} h(z_t) z_t^l \left[\sum_{m=1}^{n} y_{im} \frac{g(y_m | z_t, x_m)}{f(y_m | x_m)} - \sum_{m=1}^{n} \frac{g(y_m | z_t, x_m)}{f(y_m | x_m)} \pi_i(z_t, x_m) \right]$$

$$= \sum_{t=1}^{k} z_t^l [r_{1it} - r_{2it}], \qquad l=0,1 \qquad (8)$$

and

$$\frac{\partial L}{\partial \beta_{ij}} = \sum_{t=1}^{k} h(z_t) \left[\sum_{m=1}^{n} y_{im} \frac{g(y_m | z_t, x_m)}{f(y_m | x_m)} x_{jm} - \sum_{m=1}^{n} \frac{g(y_m | z_t, x_m)}{f(y_m | x_m)} x_{jm} \pi_i(z_t, x_m) \right]$$

$$= \sum_{t=1}^{k} z_t^l [r_{3ijt} - r_{4ijt}], \qquad j=1,...,r \qquad (9)$$

where

$$r_{1it} = h(z_t) \sum_{m=1}^{n} y_{im} g(y_m | z_t, x_m)/f(y_m | x_m) =$$
$$= \sum_{m=1}^{n} y_{im} h(z_t | y_m, x_m) \qquad (10)$$

$$r_{2it} = h(z_t) \sum_{m=1}^{n} \pi_i(z_t, x_m) g(y_m | z_t, x_m)/f(y_m | x_m) =$$
$$= \sum_{m=1}^{n} \pi_i(z_t, x_m) h(z_t | y_m, x_m) \qquad (11)$$

$$r_{3ijt} = h(z_t) \sum_{m=1}^{n} y_{im} x_{jm} g(y_m | z_t, x_m)/f(y_m | x_m) =$$
$$= \sum_{m=1}^{n} y_{im} x_{jm} h(z_t | y_m, x_m) \qquad (12)$$

and

$$r_{4ijt} = h(z_t) \sum_{m=1}^{n} x_{jm} \pi_i(z_t, x_m) g(y_m | z_t, x_m)/f(y_m | x_m) =$$
$$= \sum_{m=1}^{n} x_{jm} \pi_i(z_t, x_m) h(z_t | y_m, x_m) \qquad (13)$$

Similar expressions with the above can be derived for the other types of variables. By formulating the model in this way it can been seen that the derivatives of the log-likelihood with respect to the unknown parameters are very easily obtained for any distribution from the exponential family. The only information we need is the first derivative of the specific function $b_i(\theta(z,x))$.

10. LATENT VARIABLE MODELS

For Normal continuous items we get explicit formulae for all the model parameters. For binary, polytomous, Gamma and Poisson items the ML equations are non-linear equations. The non-linear equations can be solved using a Newton-Raphson iterative scheme. The scale parameter is one for all the cases but the normal and the gamma. Partial derivatives of the log-likelihood are computed for the scale parameter as well but they are not given here (for more details see Moustaki & Knott, 2000).

3.1 EM Algorithm

Let the vector of observed variables be $y' = (y_1, \ldots, y_L)$ where y_j is a vector of observed variables all of the same type but the type of variables is different for different values of j, $j = 1, \ldots, L$, where L denotes the number of different types of observed variables to be analyzed.

The maximization of the log-likelihood is done using the EM algorithm. The steps of the algorithm are defined as follows:

Step 1 Choose initial estimates for the model parameters $(\alpha_{il}, \beta_{ij}$ and the scale parameter ϕ_i, $i = 1, \ldots, p)$.

Step 2 E-step: Compute the r_{it} and r_{ijt} expressions. Those were given in section 3 for the binary case.

Step 3 M-step: Obtain improved estimates for the parameters by solving the non-linear maximum likelihood equations for Bernoulli, Multinomial, Gamma, and Poisson distributed variables and using the explicit equations for Normal distributed variables.

Step 4 Return to step 2 and continue until convergence is attained.

3.2 Goodness of fit

For the mixed latent trait model, the goodness-of fit of the latent trait model has been looked at separately for the categorical and the continuous part. For the categorical part of the model, significant information concerning the goodness-of fit of the model is found in the margins. That is, the one- two- and three-way margins of the differences between the observed and expected frequencies under the model are investigated for any large discrepancies for pairs and triples of items which will suggest that the model does not fit well for these combinations of items. Bartholomew and Tzamourani (1999) and Bartholomew, Steele, Moustaki and Galbraith (2002) used this called residual analysis for the latent trait model with binary and categorical items.

For the normal part of the model, we check the discrepancies between the sample covariance matrix and the one estimated from the model.

Instead of testing the goodness-of fit of a specified model we could alternatively use a criterion for selecting among a set of different models. This is particularly useful for the determination of the number of factors required. Sclove (1987) gives a review of some of the model selection criteria used in multivariate analysis such as those due to Akaike, Schwarz, and Kashap. These criteria take into account the value of the likelihood at the maximum likelihood solution and the number of parameters estimated.

Akaike's criterion for the determination of the order of an autoregressive model in time series has also been used for the determination of the number of factors in factor analysis, see Akaike (1987).

$$AIC = -2[\max L] + 2m \qquad (14)$$

where m is the number of model parameters. The model with the smallest AIC value is taken to be the best one.

4. Applications

In this section we use two real data sets to illustrate the methodology developed. One data set consists of mixed manifest items and the other data set consists of binary manifest items that are indicators of a single latent variable and two observed explanatory variables.

4.1 Some variables on welfare

Five welfare indicators are used here to construct a welfare or poverty index. The five items are from the Enquete sur la Consommation 1990 survey of the office federal de la statistique of Switzerland. Information is collected from 1963 households. Outliers have been removed from the data. That leaves us with 1923 households. The items are:

1. Presence of a dishwasher (1/0) [DishWasher]
2. Presence of a video recorder (1/0) [Video]
3. Presence of a car (1/0) [Car]
4. Equivalent food expenditures [FoodExp]
5. Equivalent expenditures for clothing [ClothExp]

These selected items were selected from a pool of welfare indicators. Items such as presence of *hot water, cooker, fridge, color TV* and *washing machine* are present in the majority of households and since they are expected to have small discrimination power they were removed from the scale.

The first 3 items are binary and items 4, 5 are metric. Judging from their frequency distributions, we treated item 4 as normally distributed and item 5 as gamma distributed. All the metric variables are standardized so that item 4 has mean zero and variance one and item 5 variance one. The one-factor model gave a good fit judging from the small discrepancies in the one -two- and three-way margins of the binary items and the AIC criterion. The AIC is 16546 for the one-factor model and 16544 for the two-factor model.

The parameter estimates are given in Table 10.2. From the last column of the table we see that the median individual is very likely to have a car (π_3 = 0.91) and less likely to have a dishwasher (π_1 = 0.40) and video recorder (π_2 = 0.38).

Table 10.2: Estimates and standard errors for the one-factor latent trait model

Binary	α_{i0}	α_{i1}	π_i (z=0)
Dishwasher	-0.43 (0.07)	1.49 (0.15)	0.40
Video	-0.51 (0.05)	0.74 (0.08)	0.38
Car	2.30 (0.15)	1.72 (0.18)	0.91
Normal	α_{i0}	α_{i1}	ϕ_i
FoodExp	0.00 (0.02)	0.50 (0.03)	0.75 (0.03)
Gamma	α_{i0}	α_{i1}	ϕ_i
ClothExp	-0.74 (0.02)	0.18 (0.01)	0.39 (0.01)

4.2 Bangladesh Fertility Survey

This data set has been extracted from the 1988 Bangladesh Fertility Survey (Huq & Cleland, 1990). The manifest items are all binary and they ask married women whether they could:
1. visit any part of the village/town/city alone
2. go outside town alone
3. talk to unknown man
4. go to the cinema alone
5. go shopping alone
6. go to club alone
7. participate in a political meeting
8. visit a health center or hospital alone.

We expect that these eight items are indicators of women's social independence. However, social independence might not be enough to explain the associations among those eight items. We could as well assume that these items are also indicators of the education of a woman and whether she lives in a rural or town/city area. The education variable is given in four categories: 'no school,' 'lower primary,' 'upper primary' and 'higher.' 'No school' is used as a reference category. The variable place of residency is given in two categories: 'rural' against 'town/city.' 'Rural' is the reference category. The results of the one-factor model are given in Table 10.3. The one-factor model gives a very good fit judging from the fit on the one- two- and three-way margins.

The estimated factor coefficients given under α_{i1} are all positive and show a strong relationship between the manifest items and the latent variable. The estimated coefficients β_{i1} show that for manifest items 4-8, a woman who lives in a town or a city is more likely to answer "yes" to those items. From the estimated coefficients for the education variable, we see that the higher the education of a woman the more likely it is for her to answer "yes" to all eight manifest items and so the higher her social independence.

5. Conclusion

The methodology developed provides a general framework for estimating the model parameters of a latent variable model with observed variables of any type from the exponential family with covariates.

A topic that was not discussed in this chapter due to the limited space is that of scoring individuals/response patterns on the identified latent dimensions. General results have been derived for the generalized model and can be found in Moustaki and Knott (2000).

The methods used to examine the goodness-of-fit of the model do not measure the overall fit of the model. However, the chi-square value computed for the one- two- and three-way margins of the binary and polytomous items is an indication of where the model might not fit well. In addition, the AIC is a way of comparing the one factor model with the two- factor model.

In this chapter, we explored the possibility of adding the direct effect of covariates on the manifest items. One could also consider modelling the effect of covariates on the latent variables (indirect effect). To decide which explanatory variables should be used for direct or indirect effects is an important topic in this area of research and very difficult to answer. A recent paper by Fayers and Hand (2002) deals with that research question without reaching any definite answers.

Table 10.3: Fertility survey: estimates for the one-factor latent trait model

Binary	α_{i0}	A_{i1}
Village/town/city	2.10	2.52
Outside town alone	-1.80	2.24
Talk to unknown man	1.35	1.41
Cinema alone	-1.54	2.70
Shopping alone	-6.10	3.41
Club alone	-5.45	3.94
Political meeting	-12.8	6.84
Health center	-5.20	3.27

Covariates	β_{i1} town/city	β_{i2} lower	β_{i3} upper	β_{i4} higher
Village/town/city	0.01	0.28	-0.06	0.96
Outside town alone	0.03	0.39	0.57	2.07
Talk to unknown man	-0.003	0.42	0.74	1.48
Cinema alone	1.53	0.29	0.76	2.69
Shopping alone	1.72	0.92	0.97	3.60
Club alone	1.04	0.006	1.35	3.63
Political meeting	2.22	1.21	2.46	6.32
Health center	1.91	0.55	0.51	3.66

6. References

Akaike, H. (1987). Factor analysis and AIC. *Psychometrika, 52*, 317-332.

Arminger, G. and Kusters, U. (1988). Latent trait models with indicators of mixed measurement level. In R. Langeheine and J. Rost (Eds.), *Latent trait and latent class models*. New York: Plenum press.

Bartholomew, D. J. and Knott, M. (1999). *Latent Variable Models and Factor Analysis* (2nd ed. London: Arnold. Kendall's Library of Statistics 7.

Bartholomew, D.J, Steele, F., Moustaki, I. and Galbraith, J. (2002). *The Analysis and Interpretation of Multivariate Data for Social Scientists*. Chapman and Hall/CRC.

Bartholomew, D. J. and Tzamourani, P. (1999). The goodness-of fit of latent trait models in attitude measurement. *Sociological Methods and Research, 27*, 525-546.

Bock, R. D. and Aitkin, M. (1981). Marginal maximum likelihood estimation of item parameters: application of EM algorithm. *Psychometrika, 46*, 443-459.

Fayers, P.M. and Hand, D.J. (2002). Causal variables, indicator variables and measurement scales: an example from quality of life. *Journal of the Royal Statistical Society, Series A*, 165, 233-261.

Huq, N. M. and Cleland, J. (1990). *Bangladesh fertility survey, 1989. Technical report*, Dhaka: National Institute of Population Research and Training (NIPORT).

Jöreskog, K. G. and Sörbom, D. (1993). *LISREL 8 - User's reference guide*. Chicago: Scientific Software, Inc.

Lawley, D. N. and Maxwell, A. E. (1971). *Factor Analysis as a Statistical Method*. London: Butterworth.

Lee, S.-Y., Poon, W.-Y. and Bender, P. (1992) . Structural equation models with continuous and polytomous variables. *Psychometrika, 57*, 89-105.

McCullagh, P. and Nelder, J. (1989). *Generalized Linear Models* (2nd ed.). London: Chapman & Hall.

Moustaki, I. (1996). A latent trait and a latent class model for mixed observed variables. *British Journal of Mathematical and Statistical Psychology, 49*, 313-334.

Moustaki, I. (2002a). GENLAT: A computer program for fitting a one- or two-factor latent variable model to categorical, metric and mixed observed items with missing values. Technical report, Statistics Department, London School of Economics and Political Science.

Moustaki, I. (2002b) A general class of latent variable models for ordinal manifest variables with covariate effects on the manifest and latent variables (under revision).

Moustaki, I. and Knott, M. (2000). Generalized latent trait models. *Psychometrika, 65*, 391-411.

Muthen, B. (1984). A general structural equation model with dichotomous, ordered categorical and continuous latent variables indicators. *Psychometrika, 49*, 115-132.

Nelder, J. and Wedderburn, R. (1972). Generalized linear models. *Journal of the Royal Statistical Society, Series A, 135*, 370-384.

Sammel, M., Ryan, L. and Legler, J. (1997). Latent variable models for mixed discrete and continuous outcomes. *Journal of the Royal Statistical Society, Series B, 59*, 667-678.

Sclove, S. (1987). Application of model-selection criteria to some problems in multivariate analysis. *Psychometrika, 52*, 333-343.

Received: July 1999. Revised: June 2002.
The paper was submitted while I. Moustaki was a Lecturer at the Department of Statistics at the London School of Economics and Political Science, UK.

11
SOME NEW ELLIPTICAL DISTRIBUTIONS

Saralees Nadarajah
Department of Mathematics
University of South Florida, Tampa, Florida

Samuel Kotz
Department of Engineering Management and Systems Engineering
George Washington University, Washington DC

1. Introduction

Multivariate elliptical distributions have received special attention in the last decade. The book by Fang et al. (1990) provides a rather detailed analysis of these distributions. Briefly, an n-dimensional random vector $\mathbf{Z}=(Z_1, \ldots, Z_n)$ is said to have an elliptically contoured distribution if its joint density takes the form:

$$f(\mathbf{z}) = |\mathbf{\Sigma}|^{-\frac{1}{2}} g((\mathbf{z}-\mathbf{\mu})^T \mathbf{\Sigma}^{-1}(\mathbf{z}-\mathbf{\mu})), \tag{1}$$

where $g(\cdot)$ is a scale function referred to as the *density generator*, $\mathbf{\Sigma}$ is an n×n constant matrix of the structure $\mathbf{A}\mathbf{A}^T$ and $\mathbf{\mu}$ is an n×1 vector. Setting

$$g(x) = \frac{\exp(-x/2)}{(2\pi)^{n/2}},$$

we arrive at the familiar multivariate n-dimensional normal distribution. Setting

$$g(x) = \frac{\Gamma(n/2)h(x)}{2\pi^{n/2} \int_0^\infty y^{n-1} h(y^2) dy},$$

where

$$h(y) = y^{N-1}\exp(-ry^s), \ r > 0, s > 0, N > 0, \tag{2}$$

we obtain the so-called symmetric Kotz type distribution (c.f. Kotz, 1975). This distribution has been studied by Fang et al. (1990), Iyengar and Tong (1989), Kotz and Ostrovskii (1994) and Streit (1991) among others. Our observation is that (2) can be viewed as the density of a Weibull-type distribution which is also known in the theory of extreme value distributions as the Type III extreme value distribution. This observation suggests that it is of interest to investigate the structure and properties of bivariate elliptical distributions with the generators corresponding to the other types of extreme value distributions; namely, the one corresponding to the Fréchet or Type II distribution

$$h(y) = y^{N-1}\exp(-ry^s), \quad r > 0, s < 0, N < 0, \qquad (3)$$

and to the Gumbel or Type I distribution

$$h(y) = \exp(-\alpha y)\exp\{-b\exp(-\alpha y)\}, \quad \alpha > 0, b > 0. \qquad (4)$$

In this chapter we derive the extremal elliptical distributions corresponding to (3)-(4) and study their structural properties. We also derive several hitherto unknown properties for the Kotz type (Weibull-type) distribution and provide a comparison of structural properties of all three extremal elliptical distributions by means of graphical illustrations.

Throughout we shall restrict attention to bivariate elliptical distributions (n=2). Without loss of generality, we shall assume that $\mu=0$ and

$$\Sigma = \begin{pmatrix} 1 & \rho \\ \rho & 1 \end{pmatrix}, \quad -1 < \rho < 1.$$

Then the joint density (1) becomes

$$f(z_1, z_2) = \frac{1}{\sqrt{1-\rho^2}} g\left(\frac{z_1^2 + z_2^2 - 2\rho z_1 z_2}{1-\rho^2}\right), \qquad (5)$$

where $g(\cdot)$ is an arbitrary generator. The marginal pdf is

$$q_g(z) = \int_{z^2}^{\infty} (x - z^2)^{-1/2} g(x) dx \qquad (6)$$

(Fang et al., 1998). Note that $q_g(z) = q_g(-z)$ for $z > 0$. By Theorems 2.7-2.8 in Fang et al. (1990), the moments associated with (5) are

$$E(Z_j) = 0, j = 1,2,$$

$$\text{Var}(Z_j) = \frac{D_1}{2}, j = 1,2,$$

$$\text{Cov}(Z_1, Z_2) = \frac{D_1 \rho}{2}$$

and $E\left(Z_1^{2t_1} Z_2^{2t_2}\right) = \pi^{-1} D_{t_1+t_2} B\left(\frac{1}{2}+t_1, \frac{1}{2}+t_2\right),$

where $t_1 \geq 1, t_2 \geq 1$ are integers

$$D_t = \pi \int_0^\infty x^t g(x) dx \qquad (7)$$

and

$$B(\alpha, \beta) = \frac{\Gamma(\alpha)\Gamma(\beta)}{\Gamma(\alpha+\beta)}.$$

By equation (2.11) in Fang et al. (1990), the characteristic function of (5) is

$$\psi(u_1, u_2) = E(\exp\{i(u_1 Z_1 + u_2 Z_2)\}) = \varphi\left(u_1^2 + u_2^2 + 2\rho u_1 u_2\right) \qquad (8)$$

where ϕ satisfies

$$\varphi(u^2) = E(\exp(iuZ_1)) = E(\exp(iuZ_2)).$$

Throughout Sections 2 and 3 of this chapter, we shall make use of the hypergeometric function defined by

$$pFq(\alpha_1, \alpha_2, \ldots, \alpha_p; \beta_1, \beta_2, \ldots, \beta_q; x) = \sum_{k=0}^{\infty} \frac{(\alpha_1)_k (\alpha_2)_k \ldots (\alpha_p)_k}{(\beta_1)_k (\beta_2)_k \ldots (\beta_q)_k} \frac{x^k}{k!}, \qquad (9)$$

See Section 9.1 of Gradshteyn and Ryzhik (1995) for details.

2. Kotz-Type (Weibull-Type) Elliptical Distribution

2.1 Joint Density

For the generator given by (2),

$$\int_0^\infty h(y) dy = \int_0^\infty y^{N-1} \exp(-ry^s) dy = \frac{\Gamma(N/s)}{sr^{N/s}}$$

and thus the corresponding joint density is

$$f(z_1, z_2) = \frac{sr^{N/s}\left(z_1^2 + z_2^2 - 2\rho z_1 z_2\right)^{N-1}}{\pi \Gamma(N/s)\left(1-\rho^2\right)^{N-1/2}} \exp\left\{-r\left(\frac{z_1^2 + z_2^2 - 2\rho z_1 z_2}{1-\rho^2}\right)^s\right\}. \qquad (10)$$

When $N = 1$, $s = 1$ and $r = 1/2$, this distribution reduces to a bivariate normal distribution. When $s = 1$, this is the original Kotz distribution introduced in Kotz (1975).

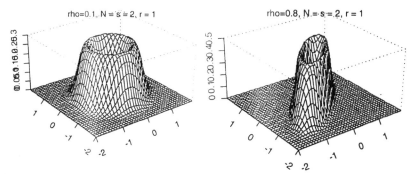

Figure 11.1: Contours of Weibull-type Elliptical Density for $\rho = 0.1$ and 0.8

The structure of the joint density (10) is illustrated in Figure 11.1 for $\rho = 0.1$ and $\rho = 0.8$ when $N = s = 2$, $r = 1$.

2.2 Marginal PDF

Let Ψ denote the degenerate hypergeometric function given by

$$\Psi(\alpha,\beta;x) = \frac{\Gamma(1-\beta)}{\Gamma(\alpha-\beta+1)} {}_1F_1(\alpha;\beta;x) + \frac{\Gamma(\beta-1)}{\Gamma(\alpha)} x^{1-\beta} {}_1F_1(\alpha-\beta+1;2-\beta;x).$$

Then using (6), the marginal pdf of (10) is

$$q_g(z) = \frac{sr^{N/s}}{\pi\Gamma(N/s)} \int_{z^2}^{\infty} (x-z^2)^{-1/2} x^{N-1} \exp(-rx^s) dx \qquad (11)$$

$$= \frac{sr^{N/s}}{\pi\Gamma(N/s)} \sum_{k=0}^{\infty} (-1)^k \binom{-1/2}{k} z^{2k} \int_{z^2}^{\infty} x^{N-k-\frac{3}{2}} \exp(-rx^s) dx$$

$$= \frac{r^{1/(2s)}}{\pi\Gamma(N/s)} \sum_{k=0}^{\infty} (-1)^k r^{k/s} \binom{-1/2}{k} z^{2k} \int_{rz^{2s}}^{\infty} y^{\frac{N}{s}-\frac{k}{s}-\frac{1}{2s}-1} \exp(-y) dy$$

$$= \frac{r^{1/(2s)} \exp(-rz^{2s})}{\pi\Gamma(N/s)} \sum_{k=0}^{\infty} (-1)^k r^{k/s} \binom{-1/2}{k} z^{2k}$$

$$\times \Psi\left(1-\frac{N}{s}+\frac{k}{s}+\frac{1}{2s}, 1-\frac{N}{s}+\frac{k}{s}+\frac{1}{2s}; rz^{2s}\right), \tag{12}$$

where the final step follows from equation (8.351.4) in Gradshteyn and Ryzhik (1995).

2.3 Moments

In equation (7),

$$D_t = \frac{sr^{N/s}}{\Gamma(N/s)}\int_0^\infty x^{t+N-1}\exp(-rx^s)dx = r^{-t/s}\Gamma\left(\frac{N}{s}+\frac{t}{s}\right)\bigg/\Gamma\left(\frac{N}{s}\right).$$

Thus the moments associated with (10) are:

$$E(Z_j) = 0, j = 1, 2,$$

$$\text{Var}(Z_j) = \frac{r^{-1/s}}{2}\Gamma\left(\frac{N}{s}+\frac{1}{s}\right)\bigg/\Gamma\left(\frac{N}{s}\right), \; j=1,2,$$

$$\text{Cov}(Z_1, Z_2) = \frac{\rho r^{-1/s}}{2}\Gamma\left(\frac{N}{s}+\frac{1}{s}\right)\bigg/\Gamma\left(\frac{N}{s}\right)$$

and for $t_1 \geq 1, t_2 \geq 1$,

$$E(Z_1^{2t_1}Z_2^{2t_2}) = \frac{r^{-(t_1+t_2)/s}}{\pi}\Gamma\left(\frac{N}{s}+\frac{t_1}{s}+\frac{t_2}{s}\right)B\left(\frac{1}{2}+t_1, \frac{1}{2}+t_2\right)\bigg/\Gamma\left(\frac{N}{s}\right).$$

2.4 Characteristic Function

The characteristic function of (10) has been studied by several authors. Iyengar and Tong (1989) gave an explicit form of the characteristic function when $s = 1$. Streit (1991) extended this result for any $s > 1/2$. Kotz and Ostrovskii (1994) obtained the structure of the characteristic function for any $0 < s < 1/2$. Here we derive the form of the function when N is an integer and $s = 1/2$. By (8), it is sufficient to work out

$$\varphi(u^2) = \int_{-\infty}^{\infty}\exp(iux)q_g(x)dx = 2\int_0^\infty \cos(ux)q_g(x)dx$$

$$= \frac{(N-1)!r^{2N}}{(2N-1)!\pi} \sum_{k=0}^{N-1} \frac{(2k)!}{2^{k-1}(N-k-1)!(k!)^2} r^{-k} \int_o^\infty \cos(ux)x^{2N-k-1} K_{k+1}(rx)dx$$

$$= 4^{1-N} \sum_{k=0}^{N-1} \frac{(2N-2k-2)!(2k)!}{((N-k-1)!)^2 (k!)^2} {}_2F_1\left(N+\frac{1}{2}, N-k-\frac{1}{2}; \frac{1}{2}; -\frac{u^2}{r^2}\right), \quad (13)$$

where the final step follows by equation (6.699.4) in Gradshteyn and Ryzhik (1995). By equation (9.131.1) in Gradshteyn and Ryzhik (1995), we can transform

$$ {}_2F_1\left(N+\frac{1}{2}, N-k-\frac{1}{2}; \frac{1}{2}; -\frac{u^2}{r^2}\right)$$

$$= r^{4N-2k-1}(u^2+r^2)^{\frac{1}{2}+k-2N} {}_2F_1\left(-N, 1-N+k; \frac{1}{2}; -\frac{u^2}{r^2}\right).$$

Substituting this back into (13), we obtain

$$\varphi(u^2) = N! 4^{1-N} \sum_{k=0}^{N-1} \frac{(2N-2k-2)!(2k)!}{(N-k-1)!(k!)^2} \sum_{l=0}^{N-k-1} \frac{(-1)^l 4^l r^{4N-2k-2l-1} u^{2l}(u^2+r^2)^{\frac{1}{2}+k-2N}}{(N-l)!(N-k-l-1)!(2l)!}. \quad (14)$$

It is worth noting here that this expression is a finite sum of elementary functions whereas those given in Streit (1991) and Kotz and Ostrovskii (1994) are both infinite sums.

3. Fréchet-Type Elliptical Distribution

3.1 Joint Density

For the generator given by (3),

$$\int_0^\infty h(y)dy = \int_0^\infty y^{N-1} \exp(-ry^s)dy = \frac{\Gamma(N/s)}{sr^{N/s}}$$

and thus the corresponding joint density is

$$f(z_1, z_2) = \frac{-sr^{N/s}(z_1^2 + z_2^2 - 2\rho z_1 z_2)^{N-1}}{\pi \Gamma(N/s)(1-\rho^2)^{N-1/2}} \exp\left\{-r\left(\frac{z_1^2 + z_2^2 - 2\rho z_1 z_2}{1-\rho^2}\right)^s\right\}. \quad (15)$$

This is similar in form to the Kotz-type distribution but both N and s are negative (thus the structure will be different).

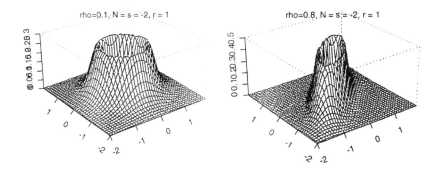

Figure 11.2: Contours of Fréchet-type Elliptical Density for $\rho = 0.1$ and 0.8

The structure is illustrated in Figure 11.2 for $\rho = 0.1$ and $\rho = 0.8$ when $N = s = -2$, $r = 1$.

3.2 Marginal PDF

Using (6), the marginal pdf of (15) is

$$q_g(z) = \frac{-sr^{N/s}}{\pi \Gamma(N/s)} \int_{z^2}^{\infty} (x - z^2)^{-1/2} x^{N-1} \exp(-rx^s) dx \qquad (16)$$

$$= \frac{-sr^{N/s}}{\pi \Gamma(N/s)|z|} \int_0^{z^{-2}} (z^{-2} - y)^{-1/2} y^{-N-\frac{1}{2}} \exp(-ry^{-s}) dy$$

$$= \frac{-sr^{N/s}}{\pi \Gamma(N/s)} \sum_{k=0}^{\infty} (-1)^k \binom{-1/2}{k} z^{2k} \int_0^{z^{-2}} y^{k-N-\frac{1}{2}} \exp(-ry^{-s}) dy$$

$$= \frac{r^{1/(2s)}}{\pi \Gamma(N/s)} \sum_{k=0}^{\infty} (-1)^k r^{k/s} \binom{-1/2}{k} z^{2k} \int_0^{rz^{2s}} y^{\frac{N}{s} - \frac{k}{s} - \frac{1}{2s} - 1} \exp(-y) dy$$

$$= \frac{2sr^{N/s} |z|^{2N-1}}{\pi \Gamma(N/s)} \sum_{k=0}^{\infty} \frac{(-1)^k}{2N - 2k - 1} \binom{-1/2}{k}$$

$$\times {}_1F_1\left(\frac{N}{s} - \frac{k}{s} - \frac{1}{2s}; \frac{N}{s} - \frac{k}{s} - \frac{1}{2s} + 1; -rz^{2s}\right), \qquad (17)$$

where the final step follows by equation (9.211.2) in Gradshteyn and Ryzhik (1995).

3.3 Moments

In equation (7),

$$D_t = \frac{-sr^{N/s}}{\Gamma(N/s)} \int_0^\infty x^{t+N-1} \exp(-rx^s) dx = r^{-t/s} \Gamma\left(\frac{N}{s} + \frac{t}{s}\right) \bigg/ \Gamma\left(\frac{N}{s}\right).$$

provided that $t<-N$. Thus the moments associated with (15) are identical to those of (10) in Section 2.3, except that we must have $N<-1$ for the variance and covariance to exist and that we must also have $t_1+t_2<-N$ for the product moment to exist.

3.4 Characteristic Function

The results in this section use the Meijer's G function defined by

$$G_{p,q}^{m,n}(x \mid \alpha_1,\ldots,\alpha_p; \beta_1,\ldots,\beta_q) = \frac{1}{2\pi i} \int_L \frac{\prod_{k=1}^m \Gamma(\beta_k - y) \prod_{k=1}^n \Gamma(1-\alpha_k+y)}{\prod_{k=m+1}^q \Gamma(1-\beta_k+y) \prod_{k=n+1}^p \Gamma(\alpha_k-y)} x^y dy$$

where "L" denotes an integration path. This function can be expressed as a finite additive combination of the hypergeometric functions, see Section 9.3 of Gradshteyn and Ryzhik (1995) for details.

Take s to be any rational number of the form $s=-m/n$ where $m\geq 1$ and $n\geq 1$ are coprime numbers. Using equation (9.34.8) in Gradshteyn and Ryzhik (1995), we can rewrite (17) as

$$q_g(z) = \frac{r^{N/s}|z|^{2N-1}}{\pi \Gamma(N/s)} \sum_{k=0}^\infty (-1)^k \binom{-1/2}{k} G_{2,1}^{1,1}\left(\frac{z^{-2s}}{r} \bigg| 1, \frac{N}{s} - \frac{k}{s} - \frac{1}{2s} + 1; \frac{N}{s} - \frac{k}{s} - \frac{1}{2s}\right).$$

Then using equation (2.24.3.3) in Prudnikov et al. (1990, Volume 3), we have

$$\varphi(u^2) = 2 \int_0^\infty \cos(ux) q_g(x) dx$$

$$= \frac{2r^{N/s}}{\pi \Gamma(N/s)} \sum_{k=0}^\infty (-1)^k \binom{-1/2}{k}$$

$$\times \int_0^\infty \cos(ux) x^{2N-1} G_{2,1}^{1,1}\left(\frac{x^{-2s}}{r} \bigg| 1, \frac{N}{s} - \frac{k}{s} - \frac{1}{2s} + 1; \frac{N}{s} - \frac{k}{s} - \frac{1}{2s}\right) dx$$

11. SOME NEW ELLIPTICAL DISTRIBUTIONS

$$= \frac{\sqrt{2}r^{N/s}(2m)^{2N}u^{-2N}}{\sqrt{mn}\,(2\pi)^{n/2}\Gamma(N/s)} \sum_{k=0}^{\infty}(-1)^k \binom{-1/2}{k} G^{n,m+n}_{2m+2n,n}\left(\frac{n^n(2m)^{2m}}{r^n u^{2m}}\bigg|\Delta(m,1-N),\right.$$

$$\left.\Delta\left(n,\left(1,\frac{N}{s}-\frac{k}{s}-\frac{1}{2s}+1\right)\right), \Delta\left(m,\frac{1}{2}-N\right); \Delta\left(n,\frac{N}{s}-\frac{k}{s}-\frac{1}{2s}\right)\right), \quad (18)$$

where

$$\Delta(l,\alpha) = \frac{\alpha}{l},\ldots,\frac{\alpha+l-1}{l}, \quad \Delta(l,(\alpha_j)) = \frac{(\alpha_j)}{l},\ldots,\frac{(\alpha_j)+l-1}{l}.$$

4. Gumbel-Type Elliptical Distribution

4.1 Joint Density

For the generator given by (4),

$$\int_0^\infty h(y)dy = \int_0^\infty \exp(-\alpha y)\exp\{-b\exp(-\alpha y)\}dy = \frac{1}{\alpha}\int_0^1 \exp(-by)dy = \frac{1-\exp(-b)}{\alpha b}$$

and thus the corresponding joint density is

$$f(z_1,z_2) = \frac{ab\left(1-\rho^2\right)^{-1/2}}{\pi(1-\exp(-b))}\exp\left[-\frac{a\left(z_1^2+z_2^2-2\rho z_1 z_2\right)}{1-\rho^2}\right]\exp\left[-b\exp\left\{-\frac{a\left(z_1^2+z_2^2-2\rho z_1 z_2\right)}{1-\rho^2}\right\}\right].$$

(19)

When $b = 0$, this distribution reduces to a bivariate normal distribution,

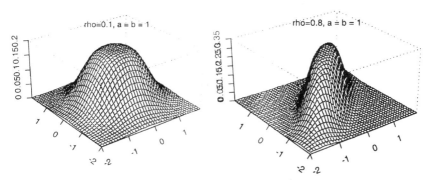

Figure 11.3: Contours of Gumbel-type Elliptical Density for $\rho = 0.1$ and 0.8

The structure of the joint density (19) is illustrated in Figure 11.3 for $\rho = 0.1$ and $\rho = 0.8$ when $\alpha = b = 1$.

4.2 Marginal PDF

Using (6), the marginal pdf of (19) is

$$q_g(z) = \frac{\alpha b}{\pi(1 - \exp(-b))} \int_{z^2}^{\infty} (x - z^2)^{-1/2} \exp(-\alpha x)\exp(-b\exp(-\alpha x))dx$$

$$= \frac{\sqrt{\alpha}\, b\exp(-\alpha z^2)}{\pi(1 - \exp(-b))} \int_0^1 (-\log y)^{-1/2} \exp(-b\exp(-\alpha z^2)y)dy$$

$$= \frac{\sqrt{\alpha}\, b\exp(-\alpha z^2)}{\pi(1 - \exp(-b))} \sum_{k=0}^{\infty} \frac{(-1)^k b^k \exp(-k\alpha z^2)}{k!} \int_0^1 (-\log y)^{-1/2} y^k dy$$

$$= \frac{\sqrt{\alpha}\, b}{\sqrt{\pi}(1 - \exp(-b))} \sum_{k=0}^{\infty} \frac{(-1)^k b^k \exp(-(k+1)\alpha z^2)}{k!\sqrt{k+1}},$$

where the final step follows by equation (4.269.4) in Gradshteyn and Ryzhik (1995).

4.3 Moments

In equation (7),

$$D_t = \frac{\alpha b}{1 - \exp(-b)} \int_0^{\infty} x^t \exp(-\alpha x)\exp(-b\exp(-\alpha x))dx$$

$$= \frac{\alpha^{-t} b}{1 - \exp(-b)} \int_0^1 (-\log y)^t \exp(-by)dy$$

$$= \frac{\alpha^{-t} b}{1 - \exp(-b)} \sum_{k=0}^{\infty} \frac{(-1)^k b^k}{k!} \int_0^1 (-\log y)^t y^k dy$$

$$= \frac{\alpha^{-t} b\, \Gamma(t+1)}{1 - \exp(-b)} \sum_{k=0}^{\infty} \frac{(-1)^k b^k}{k!(k+1)^{t+1}},$$

where the final step follows by equation (4.272.6) in Gradshteyn and Ryzhik (1995). Thus the moments associated with (19) are:

$$E(Z_j) = 0, \quad j = 1, 2,$$

$$\text{Var}(Z_j) = \frac{b}{\alpha(1-\exp(-b))} \sum_{k=0}^{\infty} \frac{(-1)^k b^k}{k!(k+1)^2}, \quad j=1,2,$$

$$\text{Cov}(Z_1, Z_2) = \frac{b\rho}{\alpha(1-\exp(-b))} \sum_{k=0}^{\infty} \frac{(-1)^k b^k}{k!(k+1)^2}$$

and for $t_1 \geq 1$, $t_2 \geq 1$,

$$E\left(Z_1^{2t_1} Z_2^{2t_2}\right) = \frac{b\Gamma(t_1+t_2+1)}{\pi\alpha^{t_1+t_2}(1-\exp(-b))} B\left(\frac{1}{2}+t_1, \frac{1}{2}+t_2\right) \sum_{k=0}^{\infty} \frac{(-1)^k b^k}{k!(k+1)^{t_1+t_2+1}}.$$

4.4 Characteristic Function

Finally, for the characteristic function we calculate ϕ as follows:

$$\varphi(u^2) = \frac{\sqrt{ab}}{\sqrt{\pi}(1-\exp(-b))} \sum_{k=0}^{\infty} \frac{(-1)^k b^k}{k!\sqrt{k+1}} \int_{-\infty}^{\infty} \exp\left[-(k+1)\alpha\left\{x^2 - \frac{iu}{(k+1)\alpha}x\right\}\right]dx$$

$$= \frac{\sqrt{ab}}{\sqrt{\pi}(1-\exp(-b))} \sum_{k=0}^{\infty} \frac{(-1)^k b^k}{k!\sqrt{k+1}} \exp\left\{-\frac{u^2}{4(k+1)\alpha}\right\} \times$$

$$\int_{-\infty}^{\infty} \exp\left[-(k+1)\alpha\left\{x - \frac{iu}{2(k+1)\alpha}\right\}^2\right] dx$$

$$= \frac{b}{1-\exp(-b)} \sum_{k=0}^{\infty} \frac{(-1)^k b^k}{(k+1)!} \exp\left\{-\frac{u^2}{4(k+1)\alpha}\right\}.$$

5. References

Fang, H. B., Fang, K. T. and Kotz, S. (1998). The meta-elliptical distributions with given marginals. Technical Report, Hong Kong Baptist University, Hong Kong.

Fang, K. T., Kotz, S. and Ng, K. W. (1990). *Symmetric Multivariate and Related Distributions.* London: Chapman and Hall.

Gradshteyn, I. S. and Ryzhik, I. M. (1995). *Table of Integrals, Series, and Products.* New York: Academic Press.

Iyengar, S. and Tong, Y. L. (1989). Convexity properties of elliptically contoured distributions. *Sankhya*, A, 51, 13-29.

Kotz, S. (1975). Multivariate distributions at a cross-road. In *Statistical Distributions in Scientific Work*, 1 G. P. Patil, S. Kotz and J. K. Ord. (eds.) Dordrecht, The Netherlands: D. Reidel Publishing Company.

Kotz, S. and Ostrovskii, I. (1994). Characteristic functions of a class of elliptical distributions. *Journal of Multivariate Analysis*, 49, 164-178.

Prudnikov, A.P., Brychkov, Y.A. and Marichev, O.I. (1990). *Integrals and Series. Volume 3: More Special Functions*. Amsterdam: Gordon and Breach Science Publishers.

Streit, F. (1991). On the characteristic functions of the Kotz type distributions. *La Societe Royale du Canada. L' Academie des Sciences. Comptes Rendus Mathematiques*, 13, 121-124.

Received: June 2000

12
EXTREME VALUE INDEX ESTIMATORS AND SMOOTHING ALTERNATIVES: A CRITICAL REVIEW

John Panaretos and Zoi Tsourti
Department of Statistics
Athens University of Economics and Business, Greece

1. Introduction

Extreme value theory is of major importance in many fields of applications where extreme values may appear and have detrimental effects. Such fields range from hydrology (Smith, 1989, Davison & Smith, 1990, Coles & Tawn, 1996, Barão & Tawn, 1999) to insurance (Beirlant et al., 1994, Mikosch, 1997, McNeil, 1997, Rootzen & Tajvidi, 1997) and finance (Danielsson & de Vries, 1997, McNeil, 1998; 1999, Embrechts et al., 1998; 1999, Embrechts, 1999). Actually, extreme value theory is a blend of a variety of applications and sophisticated mathematical results on point processes and regular varying functions.

The cornerstone of extreme value theory is Fisher-Tippet's theorem for limit laws for maxima (Fisher & Tippet, 1928). According to this theorem, if the maximum value of a distribution function (d.f.) tends (in distribution) to a non-degenerate d.f. then this limiting d.f. can *only* be the Generalized Extreme Value (GEV) distribution:

$$H_\theta(x) = H_{\gamma,\mu,\sigma}(x) = \exp\left\{-\left(1+\gamma\frac{x-\mu}{\sigma}\right)^{-1/\gamma}\right\},$$

where $1+\gamma\frac{x-\mu}{\sigma} > 0$, and $\theta = (\gamma,\mu,\sigma) \in \Re \times \Re \times \Re_+$.

A comprehensive sketch of the proof can be found in Embrechts et al. (1997). The random variable (r.v.) X (the d.f. F of X, or the distribution of X) is said to belong to the maximum domain of attraction of the extreme value distribution H_γ if there exist constants $c_n > 0$, $d_n \in \Re$ such that $c_n^{-1}(M_n - d_n) \xrightarrow{d} H_\gamma$ holds. We write $X \in MDA(H_\gamma)$ (or $F \in MDA(H_\gamma)$).

In this chapter, we deal with the estimation of the shape parameter γ known also as the *tail index or the extreme-value index*. In section 2 the general theoretical background is provided. In section 3, several existing estimators for γ are presented, while, in section 4, some smoothing methods on specific estimators are given and extended to other estimators, too. In section 5, techniques for dealing with the issue of choosing the threshold value of k, the number of upper order statistics required for the stimationj of γ, are discussed. Finally, concluding remarks are given in section 6.

2. Modelling Approaches

A starting point for modelling the extremes of a process is based on distributional models derived from asymptotic theory. The *parametric approach* to modelling extremes is based on the assumption that the data in hand (X_1, X_2, ..., X_n) form an i.i.d. sample from an exact GEV d.f. In this case, standard statistical methodology from parametric estimation theory can be utilized in order to derive estimates of the parameters θ. In practice, this approach is adopted whenever the dataset consists of maxima of independent samples (e.g., in hydrology where we have disjoint time periods). This method is often called method of *block maxima* (initiated by Gumbel, 1958). Such techniques are discussed in DuMouchel (1983), Hosking (1985), Hosking et al. (1985), Smith (1985), Scarf (1992), Embrechts et al. (1997) and Coles and Dixon (1999). However, this approach may seem restrictive and not very realistic since the grouping of data into epochs is sometimes rather arbitrary, while by using only the block maxima, we may lose important information (some blocks may contain several among the largest observations, while other blocks may contain none). Moreover, in the case that we have a few data, block maxima cannot be actually implemented.

In this chapter, we examine another widely used approach, the so-called 'Maximum Domain of Attraction or Non-Parametric Approach' (Embrechts et al., 1997). In the present context we prefer the term '*semi-parametric*' since this term reflects the fact that we make only partial assumptions about the unknown d.f. F.

So, essentially, we are interested in the distribution of the maximum (or minimum) value. Here is the point where extreme-value theory gets involved. According to the Fisher-Tippet theorem, the limiting d.f. of the (normalized) maximum value (if it exists) is the GEV d.f. $H_\theta = H_{\gamma;\mu,\sigma}$. So, without making any assumptions about the unknown d.f. F (apart from some continuity conditions which ensure the existence of the limiting d.f.), extreme-value theory provides us with a fairly sufficient tool for describing the behavior of extremes of the distribution that the data in hand stem from. The only issue that remains to be resolved is the estimation of the parameters of the GEV d.f. $\theta = (\gamma, \mu, \sigma)$. Of these parameters, the *shape parameter* γ is the one that attracts most of the attention, since it is the parameter that determines, in general terms, the behavior of extremes.

According to extreme-value theory, these are the parameters of the GEV d.f. that the maximum value follows asymptotically. Of course, in reality, we only have a finite sample and, in any case, we cannot use only the largest observation for inference. So, the procedure followed in practice is that we assume that the asymptotic approximation is achieved for the largest k observations (where k is large but not as large as the sample size n), which we subsequently use for the estimation of the parameters. However, the choice of k is not an easy task. On the contrary, it is a very controversial issue. Many authors have suggested alternative methods for choosing k, but no method has been universally accepted.

3. Semi-Parametric Extreme-Value Index Estimators

In this section, we give the most prominent answers to the issue of parameter estimation. We mainly concentrate on the estimation of the shape parameter γ due to its (already stressed) importance. The setting on which we are working is :
Suppose that we have a sample of i.i.d r.v.'s $X_1, X_2, ..., X_n$ (where $X_{1:n} \geq X_{2:n} \geq ... \geq X_{n:n}$ are the corresponding descending order statistics) from an unknown *continuous* d.f. F. According to extreme-value theory, the normalized maximum of such a sample follows asymptotically a GEV d.f. $H_{\gamma;\mu,\sigma}$, i.e.,

$$F \in MDA(H_{\gamma,\mu,\sigma})$$

In the remainder of this section, we describe the most known suggestions to the above question of estimation of extreme-value index γ, ranging from the first contributions, of 1975, in the area to very recent modifications and new developments.

3.1 The Pickands Estimator

The Pickands estimator (Pickands, 1975), is the first suggested estimator for the parameter $\gamma \in \Re$ of GEV d.f and is given by the formula

$$\hat{\gamma}_P = \frac{1}{\ln 2} \ln \left(\frac{X_{(k/4):n} - X_{(k/2):n}}{X_{(k/2):n} - X_{k:n}} \right).$$

The original justification of Pickands's estimator was based on adopting a percentile estimation method for the differences among the upper-order statistics. A more formal justification is provided by Embrechts et al. (1997).

The properties of Pickands's estimator were mainly explored by Dekkers and de Haan (1989). They proved, under certain conditions, weak and strong consistency, as well as asymptotic normality. Consistency depends only on the behavior of k, while asymptotic normality requires more delicate conditions (2nd order conditions) on the underlying d.f. F, which are difficult to verify in practice. Still, Dekkers and de Haan (1989) have shown that these conditions hold for various known and widely-used d.f.'s (normal, gamma, GEV, exponential, uniform, Cauchy).

A particular characteristic of Pickands's estimator is the fact that the largest observation is not explicitly used in the estimation. One can argue that this makes sense since the largest observation may add too much uncertainty.

Generalizations of Pickands's estimator have been introduced sharing its virtues and rendering its problems. Innovations are related to both alternative values of the multiplicative spacing parameter '2' as well as convex combinations over different k values (Yun, 2000; Segers, 2001a).

3.2 The Hill Estimator

The most popular tail index estimator is the Hill estimator, (Hill, 1975) which, however, is restricted to the Fréchet case $\gamma > 0$. The Hill estimator is provided by the formula

$$\hat{\gamma}_H = \frac{1}{k} \sum_{i=1}^{k} \ln X_{i:n} - \ln X_{k+1:n}.$$

The *original derivation* of the Hill estimator relied on the notion of conditional maximum likelihood estimation method.

The statistical behavior and properties of the Hill estimator have been studied by many authors separately, and under diverse conditions. Weak and strong consistency as well as asymptotic normality of the Hill estimator hold under the assumption of i.i.d. data (Embrechts et al., 1997). Similar (or slightly modified) results have been derived for data with several types of dependence or some other specific structures (see for example Hsing, 1991, as well as Resnick and Stărică, 1995, 1996, and 1998).

Note that the conditions on k and d.f. F that ensure the consistency and asymptotic normality of the Hill estimator are the same as those imposed for the Pickands estimator. Such conditions have been discussed by many authors, such as Davis and Resnick (1984), Haeusler and Teugels (1985), de Haan and Resnick (1998).

Though the Hill estimator has the apparent disadvantage that is restricted to the case $\gamma>0$, it has been widely used in practice and extensively studied by statisticians. Its popularity is partly due to its simplicity and partly due to the fact that in most of the cases where extreme-value analysis is called for, we have long-tailed d.f.'s (i.e., $\gamma>0$). However, its popularity generated a tempting problem, namely to try to extend the Hill estimator to the general case $\gamma \in \Re$. Such an attempt, led Beirlant et al. (1996) to the so-called adapted Hill estimator, which is applicable for $\gamma \in \Re$. Recent generalizations of the Hill estimator for $\gamma \in \Re$ are presented by Gomes and Martins (2001).

3.3 The Moment Estimator

Another estimator that can be considered as an adaptation of the Hill estimator, in order to obtain consistency for all $\gamma \in \Re$, has been proposed by Dekkers et al. (1989). This is the moment estimator, given by

$$\hat{\gamma}_M = M_1 + 1 - \frac{1}{2}\left(1 - \frac{(M_1)^2}{M_2}\right)^{-1}, \quad \text{where} \quad M_j \equiv \frac{1}{k}\sum_{i=1}^{k}\left(\ln X_{i:n} - \ln X_{(k+1):n}\right)^j,$$

j=1, 2.

Weak and strong consistency, as well as asymptotic normality of the moment estimator have been proven by its creators Dekkers et al. (1989).

3.4 The Moment-Ratio Estimator

Concentrating on cases where $\gamma > 0$, the main disadvantage of the Hill estimator is that it can be severely biased, depending on the 2nd order behavior of the underlying d.f. F. Based on an asymptotic 2nd order expansion of the d.f. F, from which one gets the bias of the Hill estimator, Danielsson et al. (1996) proposed the moment-ratio estimator defined by

$$\hat{\gamma}_{MR} = \frac{1}{2} \cdot \frac{M_2}{M_1}.$$

They proved that $\hat{\gamma}_{MR}$ has a lower asymptotic square bias than the Hill estimator (when evaluated at the same threshold, i.e., for the same k), though the convergence rates are the same.

3.5 Peng's and W estimators

An estimator related to the moment estimator $\hat{\gamma}_M$ is Peng's estimator, suggested by Deheuvels et al. (1997):

$$\hat{\gamma}_L = \frac{M_2}{2M_1} + 1 - \frac{1}{2}\left(1 - \frac{(M_1)^2}{M_2}\right)^{-1}.$$

This estimator has been designed to somewhat reduce the bias of the moment estimator.

Another related estimator suggested by the same authors is the W estimator

$$\hat{\gamma}_W = 1 - \frac{1}{2}\left(1 - \frac{(L_1)^2}{L_2}\right)^{-1}, \quad \text{where} \quad L_j \equiv \frac{1}{k}\sum_{i=1}^{k}\left(X_{i:n} - X_{(k+1):n}\right)^j, \text{ j=1, 2.}$$

As Deheuvels et al. (1997) mentioned, $\hat{\gamma}_L$ is consistent for any $\gamma \in \Re$ (under the usual conditions), while $\hat{\gamma}_W$ is consistent only for $\gamma < 1/2$. Moreover, under appropriate conditions on F and k(n), $\hat{\gamma}_L$ is asymptotically normal. Normality holds for $\hat{\gamma}_W$ only for $\gamma < 1/4$.

3.6 Estimators based on QQ plots

One of the approaches concerning Hill's derivation is the 'QQ-plot' approach (Beirlant et al., 1996). According to this, the Hill estimator is approximately the slope of the line fitted to the upper tail of Pareto QQ plot. A more precise estimator, under this approach, has been suggested by Kratz and Resnick (1996), who derived the following estimator of γ:

$$\hat{\gamma}_{qq} = \frac{\sum_{i=1}^{k} \ln\frac{i}{k+1} \left\{ \sum_{j=1}^{k} \ln X_{j:n} - k \ln X_{i:n} \right\}}{k \sum_{i=1}^{k} \left(\ln\frac{i}{k+1} \right)^2 - \left(\sum_{i=1}^{k} \ln\frac{i}{k+1} \right)^2}.$$

The authors proved weak consistency and asymptotic normality of the qq-estimator (under conditions similar to the ones imposed for the Hill estimator). However, the asymptotic variance of the qq-estimator is twice the asymptotic variance of the Hill estimator, while similar conclusions are drawn from simulations of small samples. On the other hand, one of the advantages of the qq-estimator over the Hill estimator is that the residuals (of the Pareto plot) contain information which potentially can be utilized to confront the bias in the estimates when the approximation is not exactly valid.

A further enhancement of this approach (estimation of γ based on Pareto QQ plot) is presented by Beirlant et al. (1999). They suggest the incorporation, in the estimation, of the covariance structure of the order statistics involved. This leads to a regression model formulation, from which a new estimator of γ can be constructed. This estimator is proved to be particularly useful in the case of bias of the standard estimators.

3.7 Estimators based on Mean Excess Plots

A graphical tool for assessing the behavior of a d.f. F is the mean excess function (MEF). The limit behavior of MEF of a distribution gives important information on the tail of that distribution function (Beirlant et al., 1995). MEF's and the corresponding mean excess plots (MEP's), are widely used in the first exploratory step of extreme-value analysis, while they also play an important role in the more systematic steps of tail index and large quantiles estimation. MEF is essentially the expected value of excesses over a threshold value u. The formal definition of MEF (Beirlant et al., 1996) is as follows:

Let X be a positive r.v. X with d.f. F and with finite first moment. Then MEF of X is

$$e(u) = E(X - u | X > u) = \frac{1}{\overline{F}(u)} \int_{u}^{x_F} \overline{F}(y) dy, \quad \text{for all } u>0.$$

The corresponding MEP is the plot of points $\{u, e(u), \text{for all } u > 0\}$.
The empirical counterpart of MEF based on sample $(X_1, X_2, ..., X_n)$, is

$$\hat{e}(u) = \frac{\sum_{i=1}^{n}(X_i - u)\mathbf{1}_{(u,\infty)}(X_i)}{\sum_{i=1}^{n}\mathbf{1}_{(u,\infty)}(X_i)}, \text{ where } \mathbf{1}_{(u,\infty)}(x) = 1 \text{ if } x > u, 0 \text{ otherwise.}$$

Usually, the MEP is evaluated at the points. In that case, MEF takes the form

$$E_k = \hat{e}(X_{(k+1):n}) = \frac{1}{k}\sum_{i=1}^{k} X_{i:n} - X_{(k+1):n}, \quad k=1,..., n.$$

If $X \in \text{MDA}(H_\gamma)$, $\gamma>0$, then it's easy to show that

$$e_{\ln X}(\ln u) = E(\ln X - \ln u | X > u) \to \gamma, \text{ as } u \to \infty.$$

Intuitively, this suggests that if the MEF of the logarithmic-transformed data is ultimately constant, then $X \in \text{MDA}(H_\gamma)$, $\gamma>0$ and the values of MEF converge to the true value γ.

Replacing u, in the above relation, by a high quantile $Q\left(1-\frac{k}{n}\right)$, or empirically by $X_{(k+1):n}$, we find that the estimator $e_{\ln X}(X_{(k+1):n})$ will be a consistent estimator of γ in case $X \in \text{MDA}(H_\gamma)$, $\gamma>0$. This holds when $k/n \to 0$ as $n \to \infty$. Notice that the empirical counterpart of $e_{\ln X}(X_{(k+1):n})$ is the well-known Hill estimator.

3.8 Kernel Estimators

Extreme-value theory dictates that if $F \in \text{MDA}(H_{\gamma;\mu,\sigma})$, $\gamma > 0$, then it holds that $F^{\leftarrow}(1-x) \in RV_{-\gamma}$, where $F^{\leftarrow}(.)$ is the generalized inverse (quantile) function corresponding to d.f. F. Csörgő et al. (1985) showed that for 'suitable' kernel functions K, it holds that

$$\int_0^{1/\lambda} \{\ln F^{\leftarrow}(1-u\lambda)d\{uK(u)\}\} \to \gamma, \text{ as } \lambda \to 0.$$

Substituting $F^{\leftarrow}$ by its empirical counterpart $F_n^{\leftarrow}$ (which is a consistent estimator of $F^{\leftarrow}$), they propose

$$\hat{\gamma}_{\text{Kernel}} = \left(\int_0^{1/\lambda} K(u)du\right)^{-1} \left(\sum_{j=1}^{n} \frac{j}{n\lambda} K\left(\frac{j}{n\lambda}\right)\{\ln X_{j:n} - \ln X_{(j+1):n}\}\right)$$

as an estimator of γ, where $\lambda = \lambda(n)$ is a bandwidth parameter, and K is a kernel function satisfying specific conditions. Under these conditions the authors prove asymptotic consistency and normality of the derived estimator. A more general class of kernel-type estimators for $\gamma \in \Re$ is given by Groeneboom et al. (2001).

3.9 The 'k-records' Estimator

A statistical notion that is closely related to extreme-value analysis is that of records, or, more generally, k-records. The k-record values (for definition see Nagaraja, 1988) are themselves revealing the extremal behavior of the d.f. F, so they can also be used to assess the extreme-value parameter $\gamma \in \Re$. Berred (1995) constructed the estimator:

$$\hat{\gamma}_{rec} = \ln \frac{X^{(k)}(n) - X^{(k)}(n-k)}{X^{(k)}(n-k) - X^{(k)}(n-2k)}.$$

Under the usual conditions on k(n) (though notice that now the meaning of k(n) is different than before), he proves weak and strong consistency of $\hat{\gamma}_{rec}$ while, by imposing 2nd order conditions on F (similar to the previous cases), he also shows asymptotic normality of $\hat{\gamma}_{rec}$.

3.10 Other Semi-Parametric Estimation Methods

Up to now, we have described analytically the best-known semi-parametric methods of estimation of parameter γ (extreme-value index) of $H_{\gamma;\mu,\sigma}$. Still, there is a vast literature on alternatives estimation methods. The applicability of extreme-value analysis on a variety of different fields led scientists with different background to work on this subject and consequently derive many and different estimators. The Pickands, Hill and, recently, the moment estimators, continue to be the basis. If nothing else, most of the other proposed estimators constitute efforts to render some of the disadvantages of these three basic estimators, while others aim to generalize the framework of these. In the sequel, we present a number of such methods.

As one may notice, apart from estimators applicable for any $\gamma \in \Re$, estimation techniques have been developed valid only for a specific range of values of γ. This is due to the fact that H_γ, for γ in a specific range, may lead to families of d.f.'s of special interest. The most typical types are estimation methods for $\gamma > 0$ which correspond to d.f.'s with regularly varying tails (here the Hill estimator is included). Moreover, estimators for $\gamma \in (0,1/2)$ are of particular interst since H_γ, $\gamma \in (0,1/2)$ represents α-stable distributions ($\gamma=1/\alpha$).

Estimators for the index $\gamma > 0$, have also been proposed by Hall (1982), Feuerverger and Hall (1999), and Segers (2001b). More restricted estimation techniques for α-stable d.f.'s are described in Mittnik et al. (1998) as well as in Kogon and Williams (1998). Sometimes, the interest of authors is focused merely on the estimation of large quantiles, which in any case is what really matters in practical situations. Such estimators have been proposed by Davis and Resnick (1984) (for $\gamma > 0$) and Boos (1984) (for $\gamma = 0$).

Under certain conditions on the 2nd order behavior of the underlying distribution the error of the Hill estimator consists of two components: the systematic bias and a stochastic error. These quantities are functions of unknown parameters, prohibiting their determination and, thus, the correction of the Hill estimator. Hall (1990), suggested the use of bootstrap resamples of small size for computing a series of values of γ to estimate its bias. This approach has been further explored and extended by Pictet et al. (1998). Furthermore, they developed a jackknife algorithm for the assessment of the stochastic error component of the Hill estimator. The bootstrap (jackknife) methodology in estimation of the extreme value index has also been discussed by Gomes et al. (2000), where generalized jackknife estimators are presented as affined combinations of Hill estimators. As the authors mention, this methodology could also be applied to other classical extreme value index estimators.

3.11 Theoretical Comparison of Estimators

So far, we have mentioned several alternative estimators for the extreme-value index γ. All of these estimators share some common desirable properties, such as weak consistency and asymptotic normality (though these properties may hold under slightly different, unverifiable in any case, conditions on F and for different ranges of the parameter γ). On the other hand, simulation studies or applications on real data can end up in large differences among these estimators. In any case, there is no 'uniformly better' estimator (i.e., an estimator that is best under all circumstances). Of course, Hill, Pickands and moment estimators are the most popular ones. This could be partly due to the fact that they are the oldest ones. The rest of the existing estimators will be introduced later. Actually, most of these have been introduced as alternatives to Hill, Pickands or moment estimators and some of them have been proven to be superior in some cases. In the literature, there are some comparison studies of extreme-value index estimators (either theoretically or via Monte-Carlo methods), such as those by Rosen and Weissman (1996), Deheuvels et al. (1997), Pictet et al. (1998), and Groeneboom et al. (2001). Still, these studies are confined to a small number of estimators. Moreover, most of the authors that introduce a new estimator compare it with some of the standard estimators (Hill, Pickands, Moment).

3.12 An Alternative Approach: The Peaks Over Threshold Estimation Method

All the previously discussed semi-parametric estimation methods were based on the notion of maximum domains of attraction of the generalized extreme-value d.f. Still, further results in extreme-value theory describe the behavior of large observations that exceed high thresholds, and these are the results which lend themselves to the so-called 'Peaks Over Threshold' (POT, in short) models. The distribution which comes to the fore in this case is the generalized Pareto distribution (GPD). Thus, the estimation of the extreme-value parameter γ or the large quantiles of the underlying d.f.'s can be alternatively estimated via the GPD instead of the generalized extreme-value distribution.

The GPD can be fitted to data consisting of excesses of high thresholds by a variety of methods including the maximum likelihood method (ML) and the method of probability weighted moments (PWM). MLEs must be derived numerically because the minimal sufficient statistics for the GPD are the order statistics and there is no obvious simplification of the non-linear likelihood equation. Grimshaw (1993) provides an algorithm for estimating the MLEs for GPD. ML and PWM methods have been compared for data of GPD both theoretically and in simulation studies by Hosking and Wallis (1987) and Rootzén and Tajvidi (1997). A graphical method of estimation (Davison & Smith, 1990) is also suggested

Here, an important practical problem is the choice of the level u of the excesses. This is analogous to the problem of choosing k (number of upper order statistics) in the previous estimators. There are theoretical suggestions on how to do this, based on a compromise between bias and variance − a higher level can be expected to give less bias, but instead gives fewer excesses, and hence a higher variance. However, these suggestions don't quite solve the problem in practice. Practical aid can be provided by QQ plots, mean excess plots and experiments with different levels u. If the model produces very different results for different choices of u, the results obviously should be viewed with more caution (Rootzén & Tajvidi, 1997).

4. Smoothing and Robustifying Procedures for Semi-Parametric Extreme-Value Index Estimators

In the previous section, we presented a series of (semi-parametric) estimators for the extreme value index γ. Still, one of the most serious objections one could raise against these methods is their sensitivity towards the choice of k (number of upper order statistics used in the estimation). The well-known phenomenon of bias-variance trade-off turns out to be unresolved, and choosing k seems to be more of an art than a science.

Some refinements of these estimators have been proposed, in an effort to produce unbiased estimators even when a large number of upper order statistics is used in the estimation (see, for example, Peng, 1998, or Drees, 1996). In the next section we present a different approach on this issue. We go back to elementary notions of extreme-value theory, and statistical analysis in general, and try to explore methods to render (at least partially) this problem. The procedures we use are (i) smoothing techniques and (ii) robustifying techniques.

4.1 Smoothing Extreme-Value Index Estimators

The essence of semi-parametric estimators of extreme-value index γ, is that we use information of only the most extreme observations in order to make inference about the behavior of the maximum of a d.f. An exploratory way to subjectively choose the number k is based on the plot of the estimator $\hat{\gamma}(k)$ versus k. A stable region of the plot indicates a valid value for the estimator. The need for a stable region results from adapting theoretical limit theorems which are proved subject to the conditions $k(n) \to \infty$ and $k(n)/n \to 0$. However, the search for a stable region in the plot is a standard but problematic and ill-defined practice. Since extreme events by definition are rare, there is only little data (few observations) that can be utilized and this inevitably involves an added statistical uncertainty. Thus, sparseness of large observations and the unexpectedly large differences between them, lead to a high volatility of the part of the plot that we are interested in and makes the choice of k very difficult. That is, the practical use of the estimator on real data is hampered by the high volatility of the plot and bias problems and it is often the case that volatility of the plot prevents a clear choice of k. A possible solution would be to smooth 'somehow' the estimates with respect to the choice of k (i.e., make it more insensitive to the choice of k), leading to a more stable plot and a more reliable estimate of γ. Such a method was proposed by Resnick and Stărică (1997, 1999) for smoothing Hill and moment estimators, respectively.

4.1.1 Smoothing the Hill Estimator

Resnick and Stărică (1997) proposed a simple averaging technique that reduces the volatility of the Hill-plot. The smoothing procedure consists of averaging the Hill estimator values corresponding to different values of the order statistics p. The formula of the proposed averaged-Hill estimator is :

$$av\hat{\gamma}_H(k) = \frac{1}{k-[ku]} \sum_{p=[ku]+1}^{k} \hat{\gamma}_H(p) ,$$

where u<1, and [x] denotes the smallest integer greater than or equal to x.

The authors proved that through averaging (using the above formula), the variance of the Hill estimator can be considerably reduced and the volatility of the plot tamed. The smoothed graph has a narrower range over its stable regime, with less sensitivity to the value of k. This fact diminishes the importance of selecting the optimal k. The smoothing techniques make no (additional) unrealistic or uncheckable assumptions and are always available to complement the Hill plot. Obviously, when considerable bias is present, the averaging technique offers no improvement.

Resnick and Stărică (1997) derived the adequacy (consistency and asymptotic normality) of the averaged-Hill estimator, as well as its improvement over the Hill estimator (smaller asymptotic variance). Since the asymptotic variance is an increasing function of u, one would like to choose u as small as possible to ensure a maximum decrease in the variance. However, the choice of u is limited by the sample size. Due to the averaging, the smaller the u, the fewer the points one gets on the plot of averaged Hill. Therefore, an equilibrium should be reached between variance reduction and a comfortable number of points on the plot. This is a problem similar to the variance-bias trade-off encountered in the simple extreme-value index estimators.

4.1.2 Smoothing the Moment Estimator

Resnick and Stărică (1999) also applied their idea of smoothing to the more general moment estimator $\hat{\gamma}_M$, essentially generalizing their reasoning of smoothing the Hill estimator.

The proposed smoothing technique consists of averaging the moment estimator values corresponding to different numbers of order statistics p. The formula of the proposed averaged-moment estimator, for given $0 < u < 1$, is :

$$av\hat{\gamma}_M(k) = \frac{1}{k-[ku]} \sum_{p=[ku]+1}^{k} \hat{\gamma}_M(p) .$$

In practice, the authors suggest to take u=0.3 or u=0.5 depending on the sample size (the smaller the sample size the larger u should be).

In this case, the consequent reduction in asymptotic variance is not so profound. The authors actually showed that through averaging (using the above formula), the variance of the moment estimator is considerably reduced only in the case $\gamma < 0$. In the case $\gamma > 0$ the simple moment estimator turns out to be superior than the averaged-moment estimator. For $\gamma \approx 0$ the two moment estimators (simple and averaged) are almost equivalent. These conclusions hold asymptotically, and have been verified via a graphical comparison, since the analytic formulas of variances are rather complicated to be compared directly. A full treatment of this issue and proofs of the propositions can be found in Resnick and Stărică (1999).

4.2 Robust Estimators Based on Excess Plots

As we have previously mentioned the MEP constitutes a starting point for the estimation of extreme-value index. In practice, strong random fluctuations of the empirical MEF and of the corresponding MEP are observed, especially in the right part of the plot (i.e., for large values of u), since there we have fewer data. But this is exactly the part of plot that mostly concerns us; that is the part that theoretically informs us about the tail behavior of the underlying d.f. Consequently, the calculation of the 'ultimate' value of MEF can be largely influenced by only a few extreme outliers, which may not even be representative of the general 'trend.' The result of Drees and Reiss (1996) that the empirical

MEF is an inaccurate estimate of the Pareto MEF, and that the shape of the empirical curve heavily depends on the maximum of the sample, is striking.
In an attempt to make the procedure more robust, that is less sensitive to the strong random fluctuations of the empirical MEF at the end of the range, the following adaptations of MEF have been considered (Beirlant et al., 1996):

- Generalized Median Excess Function $M^{(p)}(k) = X_{([pk]+1):n} - X_{(k+1):n}$

(for p=0.5 we get the simple median excess function).

- Trimmed Mean Excess Function $T^{(p)}(k) = \dfrac{1}{k-[pk]} \sum_{j=[pk]+1}^{k} X_{j:n} - X_{(k+1):n}$.

The general motivations and procedures explained for the MEF and its contribution to the estimation of γ hold here as well. Thus, alternative estimators for γ>0 are :

- $\hat{\gamma}_{\text{gen med}} = \dfrac{1}{\ln(1/p)} \left(\ln X_{([pk]+1):n} - \ln X_{(k+1):n} \right)$

which for p=0.5 gives $\hat{\gamma}_{\text{med}} = \dfrac{1}{\ln(2)} \left(\ln X_{([k/2]+1):n} - \ln X_{(k+1):n} \right)$

(the consistency of this estimator is proven by Beirlant et al., 1996).

- $\hat{\gamma}_{\text{trim}} = \dfrac{1}{k-[pk]} \sum_{j=[pk]+1}^{k} \ln X_{j:n} - \ln X_{(k+1):n}$

It is worth noting that robust estimation of the tail index of a two-parameter Pareto distribution is presented by Brazauskas and Serfling (2000). The corresponding estimators are of generalized quantile type. The authors distinguish the trimmed mean and generalized median type as the most competitive trade-offs between efficiency and robustness.

5. More Formal Methods for Selecting k

In the previous sections we have presented some attempts to derive extreme-value index estimators, smooth enough, so that the plot $\{k, \hat{\gamma}(k)\}$ is an adequate tool for choosing k and consequently deciding on the estimate $\hat{\gamma}(k)$. However, such a technique will always be a subjective one and there are cases where we need a more objective solution. Actually, there are cases where we need a quick, automatic, clear-cut choice of k. So, for reasons of completeness, we present some methods for choosing k in extreme-value index estimation. Such a choice of k is, essentially, an 'optimal choice,' in the sense that we are looking for the optimal sequence k(n) that balances the variance and bias of the estimators. This optimal sequence $k_{\text{opt}}(n)$ can be determined when the underlying distribution F is known, provided that the d.f. has a second order expansion involving an extra unknown parameter. Adaptive methods for choosing k were proposed for special classes of distributions (see Beirlant et al.,

1996 and references in Resnick and Stărică, 1997). However, such second order conditions are unverifiable in practice. Still Dekkers and de Haan (1993) prove that such conditions hold for some well-known distributions (such as the Cauchy, the uniform, the exponential, and the generalized extreme-value distributions). Of course, in practice we do not know the exact analytic form of the underlying d.f. So, several approximate methods, which may additionally estimate (if needed) the 2nd order parameters, have been developed. Notice, that the methods existing in the literature are not generally applicable to any extreme-value index estimator but are designed for particular estimators in each case.

Drees and Kaufmann (1998), proposed a sequential approach to construct a consistent estimator of k that works asymptotically without any prior knowledge about the underlying d.f. Recently, a simple diagnostic method for selecting k has been suggested by Guillou and Hall (2001). They performed a sort of hypothesis testing on log-spacings by appropriately weighting them. Both of these approaches have been originally introduced for the Hill estimator, but can be extended to other extreme-value index estimators, too.

5.1 The Regression Approach

Recall that according to the graphical justification of the Hill estimator, this estimator can be derived as the estimation of the slope of a line fitted to the k upper-order statistics of our dataset. In this sense, the choice of k can be reduced to the problem of choosing an anchor point to make the linear fit optimal. In statistical practice, the most common measure of optimality is the mean square error.

In the context of the Hill estimator (for $\gamma > 0$) and the adapted Hill estimator (for $\gamma \in \Re$), Beirlant et al. (1996) propose the minimization of the asymptotic mean square error of the estimator as an appropriate optimality criterion. They have suggested using

$$\text{MSE}_{opt}(k) = \frac{1}{k}\sum_{j=1}^{k} w_{j,k}^{opt}\left(\ln X_{j:n} - \left[\ln X_{(k+1):n} + \gamma \ln \frac{k+1}{j}\right]\right)$$

as a consistent estimate (as $n \to \infty$, $k \to \infty$, $k/n \to 0$) of asymptotic mean square error of Hill estimator ($w_{j,k}^{opt}$ is a sequence of weights).

Theoretically, it would suffice to compute MSE_{opt} for every relevant value of k and look for the minimal MSE value with respect to k. Note that in the above expression neither γ (true value of extreme-value index) nor the weights $w_{j,k}^{opt}$, which depend on a parameter ρ of the 2nd order behavior of F, are known. So, Beirlant et al. (1996), propose an iterative algorithm for the search of the optimum k.

5.2 The Bootstrap Approach

Draisma et al. (1999) developed a purely sample-based method for obtaining the optimal sequence $k_{opt}(n)$. They, too, assume a second order expansion of the underlying d.f., but the second (or even the first) order parameter is not required to be known. In particular, their procedure is based on a double bootstrap. They are concerned with the more general case $\gamma \in \Re$, and their results refer to the Pickands and the moment estimators.

As before, they want to determine the value of k, $k_{opt}(n)$, minimizing the asymptotic mean square error $E_F(\hat{\gamma}(k) - \gamma)$, where $\hat{\gamma}$ refers either to the Pickands estimator $\hat{\gamma}_P$ or to the moment estimator $\hat{\gamma}_M$. However, in the above expression there are two unknown factors: the parameter γ and the d.f. F. Their idea is to replace γ by a second estimator $\hat{\gamma}_+$ (its form depending on whether we use the Pickands or the moment estimator) and F by the empirical d.f. F_n. This is determined by bootstrapping. The authors prove that minimizing the resulting expression, which can be calculated purely on the basis of the sample, still leads to the optimal sequence $k_{opt}(n)$ again via a bootstrap procedure.

The authors test their proposed bootstrap approach on various d.f.'s (such as those of the Cauchy, the generalized Pareto, and the generalized extreme-value distributions) via simulation. The general conclusion is that the bootstrap procedure gives reasonable estimates (in terms of mean square error of the extreme-value index estimator) for the sample fraction to be used. So, such a procedure takes out the subjective element of choosing k. However, even in such a procedure an element of subjectivity remains, since one has to choose the number of bootstrap replications (r) and the size of the bootstrap samples (n_1).

Similar bootstrap-based methods for selecting k have been presented by Danielsson and de Vries (1997) and Danielsson et al. (2000), confined to $\gamma > 0$, with results concerning only the Hill estimator $\hat{\gamma}_H$. Moreover, Geluk and Peng (2000) apply a 2-stage non-overlapping subsampling procedure, in order to derive the optimal sequence $k_{opt}(n)$ for an alternative tail index estimator (for $\gamma > 0$) for finite moving average time series.

6. Discussion and Open Problems

The wide collection of estimators of the extreme value index which characterizes the tails of most distributions, has been the central issue of this chapter. We presented the main approaches for the estimation of γ, with special emphasis to the semi-parametric one. In sections 3 and 4 several such estimators are provided (Hill, Moment, Pickands, among others). Some modifications of these proposed in the literature based on smoothing and robustifying procedures have also been considered since the dependence of these estimators on the very extreme observations which can display very large deviations, is one of their drawbacks.

Summing up, there is not a uniformly best estimator of the extreme-value index. On the contrary, the performance of estimators seems to depend on the distribution of data in hand. From another point of view, one could say that the performance of estimators of the extreme-value index depends on the value of the index itself. So, before proceeding to the use of any estimation formula it would be useful if we could get an idea about the range of values where the true γ lies in. This can be achieved graphically via QQ and mean excess plots. Alternatively, there exist statistical tests that tests such a hypothesis. (See, for example, Hosking, 1984, Hasofer & Wang, 1992, Alves & Gomes, 1996, and Marohn, 1998; 2000, Segers & Teugels, 2000).

However, the 'Achilles heel' of semi-parametric estimators of the extreme-value index is its dependence and sensitivity on the number k of upper order statistics used in the estimation. No hard and fast rule exists for confronting this problem. Usually, the scientist subjectively decides on the number k to use, by looking at appropriate graphs. More objective ways for doing this are based on regression or bootstrap. The bootstrap approach is a newly suggested and promising method in the field of extreme-value analysis. Another area of extreme-value index estimation where bootstrap methodology could turn out to be very useful is in the estimation (and, consequently, elimination) of the bias of extreme-value index estimators. The bias is inherent in all of these estimators, but it is not easy to be assessed theoretically because it depends on second order conditions on the underlying distribution of data, which are usually unverifiable. Bootstrap procedures could approximate the bias without making any such assumptions.

Finally, we should mention that a new promising branch of extreme-value analysis is that of multivariate extreme-value methods. One of the problems in extreme-value analysis is that, usually, one deals with few data which leads to great uncertainty. This drawback can be aleviated somehow, by the simultaneous use of more than one source of information (variables), i.e., by applying multivariate extreme-value analysis. Such an approach is attempted by Embrechts, de Haan and Huang (1999) and Caperaa and Fougeres (2001). This technique has already been applied to the field of hydrology. See, for example, de Haan and de Ronde (1998), de Haan and Sinha (1999) and Barão and Tawn (1999).

7. References

Alves, M.I.F. and Gomes, M.I. (1996). Statistical Choice of Extreme Value Domains of Attraction – A Comparative Analysis. *Communication in Statistics (A) – Theory and Methods,* 25 (4), 789-811.

Barão, M.I. and Tawn, J.A. (1999). Extremal Analysis of Short Series with Outliers: Sea-Levels and Athletic Records. *Applied Statistics,* 48 (4), 469-487.

Beirlant, J., Broniatowski, M., Teugels, J.L. and Vynckier, P. (1995). The Mean Residual Life Function at Great Age: Applications to Tail Estimation. *Journal of Statistical Planning and Inference*, 45, 21-48.

Beirlant, J., Dierckx, G., Goegebeur, Y. and Matths, G. (1999). Tail Index Estimation and an Exponential Regression Model. *Extremes*, 2(2), 177-200.

Beirlant, J., Teugels, J.L. and Vynckier, P. (1994). Extremes in Non-Life Insurance. *Extremal Value Theory and Applications*, Galambos, J., Lenchner, J. and Simiu, E. (eds.), Dordrecht, Kluwer, 489-510.

Beirlant, J., Teugels, J.L. and Vynckier, P. (1996). *Practical Analysis of Extreme Values*. Leuven University Press, Leuven.

Berred, M. (1995). K-record Values and the Extreme-Value Index. *Journal of Statistical Planning and Inference*, 45, 49-63.

Boos, D.D. (1984). Using Extreme Value Theory to Estimate Large Percentiles. *Technometrics*, 26 (1), 33-39.

Brazauskas, V. and Serfling., R. (2000). Robust Estimation of Tail Parameters for Two-Parameter Pareto and Exponential Models via generalized Quantile Statistics. *Extremes*, 3(3), 231-249.

Caperaa, P. and Fougeres, A.L. (2001). Estimation of a Bivariate Extreme Value Distribution. *Extremes*, 3(4), 311-329.

Coles, S.G. and Dixon, M.J. (1999). Likelihood-Based Inference for Extreme Value Models. *Extremes*, 2 (1), 5-23.

Coles, S.G. and Tawn, J.A. (1996). Modelling Extremes of the Areal Rainfall Process. *Journal of the Royal Statistical Society, B* 58 (2), 329-347.

Csörgő, S., Deheuvels, P. and Mason, D. (1985). Kernel Estimates of the Tail Index of a Distribution. *The Annals of Statistics*, 13 (3), 1055-1077.

Danielsson, J. and de Vries, C.G. (1997). Beyond the Sample: Extreme Quantile and Probability Estimation. *Preprint*, Erasmus University, Rotterdam.

Danielsson, J., de Haan, L., Peng, L. and de Vries, C.G. (2000). Using a Bootstrap Method to Choose the Sample Fraction in Tail Index Estimation. *Journal of Multivariate Analysis*, 76, 226-248.

Danielsson, J., Jansen, D.W. and deVries, C.G. (1996). The Method of Moment Ratio Estimator for the Tail Shape Distribution. *Communication in Statistics (A) – Theory and Methods*, 25 (4), 711-720.

Davis, R. and Resnick, S. (1984). Tail Estimates Motivated by Extreme Value Theory. *The Annals of Statistics*, 12 (4), 1467-1487.

Davison, A.C. and Smith, R.L. (1990). Models for Exceedances over High Thresholds. *Journal of the Royal Statistical Society, B* 52 (3), 393-442.

De Haan, L. and de Ronde, J. (1998). Sea and Wind: Multivariate Extremes at Work. *Extremes*, 1 (1), 7-45.

De Haan, L. and Resnick, S. (1998). On Asymptotic Normality of the Hill Estimator. *Stochastic Models*, 14, 849-867.

De Haan, L. and Sinha, A.K. (1999). Estimating the Probability of a Rare Event. *The Annals of Statistics*, 27 (2), 732-759.

Deheuvels, P., de Haan, L., Peng, L. and Pereira, T.T. (1997). Comparison of Extreme Value Index Estimators. NEPTUNE T400:EUR-09.

Dekkers, A.L.M. and de Haan, L. (1989). On the Estimation of the Extreme-Value Index and Large Quantile Estimation. *The Annals of Statistics*, 17 (4), 1795 - 1832.

Dekkers, A.L.M. and de Haan, L. (1993). Optimal Choice of Sample Fraction in Extreme-Value Estimation. *Journal of Multivariate Analysis*, 47(2), 173-195.

Dekkers, A.L.M., Einmahl, J.H.J. and de Haan, L. (1989). A Moment Estimator for the Index of an Extreme-Value Distribution. *The Annals of Statistics*, 17 (4), 1833-1855.

Draisma, G., de Haan, L., Peng, L. and Pereira, T.T. (1999). A Bootstrap-Based Method to Achieve Optimality in Estimating the Extreme-Value Index. *Extremes*, 2 (4), 367-404.

Drees, H. (1996). Refined Pickands Estimators with Bias Correction. *Communication in Statistics (A) – Theory and Methods*, 25 (4), 837-851.

Drees, H. and Kaufmann, E. (1998). Selecting the Optimal Sample Fraction in Univariate Extreme Value Estimation. *Stochastic Processes and their Applications*, 75, 149-172.

Drees, H. and Reiss, R.D. (1996). Residual Life Functionals at Great Age. *Communication in Statistics (A) – Theory and Methods,* 25 (4), 823-835.

DuMouchel, W.H. (1983). Estimating the Stable Index α in Order to Measure Tail Thickness: a Critique. *The Annals of Statistics,* 11, 1019-1031.

Embrechts, P. (1999). Extreme Value Theory in Finance and Insurance. *Preprint,* ETH.

Embrechts, P., de Haan, L. and Huang, X. (1999). Modelling Multivariate Extremes. *ETH Preprint.*

Embrechts, P., Klüppelberg, C. and Mikosch, T. (1997). *Modelling Extremal Events for Insurance and Finance.* Springer, Berlin.

Embrechts, P., Resnick, S. and Samorodnitsky, G. (1998). Living on the Edge. *RISK Magazine,* 11 (1), 96-100.

Embrechts, P., Resnick, S. and Samorodnitsky, G. (1999). Extreme Value Theory as a Risk Management Tool. *North American Actuarial Journal,* 26, 30-41.

Feuerverger, A. and Hall, P. (1999). Estimating a Tail Exponent by Modelling Departure from a Pareto Distribution. *The Annals of Statistics,* 27(2), 760-781.

Fisher, R.A. and Tippet, L.H.C. (1928). Limiting Forms of the Frequency Distribution of the Largest or Smallest Member of a Sample. *Proc. Cambridge Phil. Soc.,* 24 (2), 163-190. (in Embrechts et al. 1997).

Geluk, J.L. and Peng, L. (2000). An Adaptive Optimal Estimate of the Tail Index for MA(1) Time Series. *Statistics and Probability Letters,* 46, 217-227.

Gomes, M.I. and Martins, M.J. (2001). Generalizations of the Hill Estimator – Asymptotic versus Finite Sample Properties. *Journal of Statistical Planning and Inference,* 93, 161-180.

Gomes, M.I., Martins, M.J., Neves, M. (2000). Alternatives to a Semi-Parametric estimator of parameters of Rare Events-The Jackknife Methodology, *Extremes,* 3(3), 207-229.

Grimshaw, A. (1993). Computing Maximum Likelihood Estimates for the Generalized Pareto Distribution. *Technometrics,* 35, 185-191.

Groeneboom, P., Lopuhaa, H.P. and de Wolf, P.P. (2001). Kernel-Type estimators for the Extreme Value Index. *Annals of Statistics,* (to appear).

Guillou, A. and Hall, P. (2001). A Diagnostic for Selecting the Threshold in Extreme Value Analysis. *Journal of the Royal Statistical Society Series B,* 63 (2), 293-350.

Gumbel, E.J. (1958). *Statistics of Extremes.* Columbia University Press, New York. (in Kinnison, 1985).

Haeusler, E. and Teugels, J.L. (1985). On Asymptotic Normality of Hill's Estimator for the Exponent of Regular Variation. *The Annals of Statistics,* 13 (2), 743-756.

Hall, P. (1982). On Some Simple Estimates of an Exponent of Regular Variation. *Journal of the Royal Statistical Society,B* 44 (1), 37-42.

Hall, P. (1990). Using the Bootstrap to Estimate Mean Square Error and Select Smoothing Parameter in NonParametric Problems. *Journal of Multivariate Analysis,* 32, 177-203.

Hasofer, A.M. and Wang, Z. (1992). A Test for Extreme Value Domain of Attraction. *Journal of the American Statistical Association,* 87, 171-177.

Hill, B.M. (1975). A Simple General Approach to Inference about the Tail of a Distribution. *The Annals of Statistics,* 3 (5), 1163-1174.

Hosking, J.R.M. (1984). Testing Whether the Shape Parameter Is Zero in the Generalized Extreme-Value Distribution. *Biometrika,* 71(2), 367-374.

Hosking, J.R.M. (1985). Maximum-Likelihood Estimation of the Parameter of the Generalized Extreme-Value Distribution. *Applied Statistics,* 34, 301-310.

Hosking, J.R.M. and Wallis, J.R. (1987). Parameter and Quantile Estimation for the Generalized Pareto Distribution. *Technometrics,* 29 (3), 333-349.

Hosking, J.R.M., Wallis, J.R. and Wood, E.F. (1985). Estimation of the Generalized Extreme-Value Distribution by the Method of Probability-Weighted Moments. *Technometrics,* 27 (3), 251-261.

Hsing, T. (1991). On Tail Index Estimation Using Dependent Data. *Annals of Statistics,* 19 (3), 1547-1569.

Kogon, S.M. and Williams, D.B. (1998). Characteristic Function Based Estimation of Stable Distribution Parameters. In R. Adler et al, (eds.), "A Practical Guide to Heavy Tails: Statistical Techniques and Applications". Boston: Birkhauser.

Kratz, M. and Resnick, S.I. (1996). The QQ Estimator and Heavy Tails. *Communication in Statistics – Stochastic Models*, 12 (4), 699-724.

Marohn, F. (1998). Testing the Gumbel Hypothesis via the Pot-Method. *Extremes*, 1(2), 191-213.

Marohn, F. (2000). Testing Extreme Value Models. *Extremes*, 3(4), 363-384.

McNeil, A.J. (1997). Estimating the Tails of Loss Severity Distributions Using Extreme Value Theory. *ASTIN Bulletin*, 27, 117-137.

McNeil, A.J. (1998). On Extremes and Crashes. A short-non-technical article. *RISK*, January 1998, 99.

McNeil, A.J. (1999). Extreme Value Theory for Risk Managers. *Internal Modelling and CAD II* published by RISK Books, 93-113.

Mikosch, T. (1997). Heavy-Tailed Modelling in Insurance. *Communication in Statistics – Stochastic Models*, 13 (4), 799-815.

Mittnik, S., Paolella, M.S. and Rachev, S.T. (1998). A Tail Estimator for the Index of the Stable Paretian Distribution. *Communication in Statistics – Theory and Method*, 27 (5), 1239-1262.

Nagaraja, H.N. (1988). Record Values and Related Statistics – A Review. *Communication in Statistics – Theory and Methods*, 17 (7), 2223-2238.

Peng, L. (1998). Asymptotically Unbiased Estimators for the Extreme-Value Index. *Statistics and Probability Letters*, 38, 107-115.

Pickands, J. (1975). Statistical Inference Using Extreme order statistics. *The Annals of Statistics*, 3 (1), 119-131.

Pictet, O.V., Dacorogna, M.M. and Muller, U.A., (1998). Hill, Bootstrap and Jackknife Estimators for Heavy Tails. In R. Adler et al. (Eds.), "A Practical Guide to Heavy Tails: Statistical Techniques and Applications". Boston: Birkhauser.

Resnick, S. and Stărică, C. (1995). Consistency of Hill's Estimator for Dependent Data. *Journal of Applied Probability*, 32, 139-167.

Resnick, S. and Stărică, C. (1996). Asymptotic Behaviour of Hill's Estimator for Autoregresive Data. Preprint, Cornell University. (Available as TR1165.ps.Z at http://www.orie.cornell.edu/trlist/trlist.html.)

Resnick, S. and Stărică, C. (1997). Smoothing the Hill Estimator. *Advances in Applied Probability*, 29, 271-293.

Resnick, S. and Stărică, C. (1998). Tail Index Estimation for Dependent Data. *Annals of Applied Probability*, 8 (4), 1156-1183.

Resnick, S. and Stărică, C. (1999). Smoothing the Moment Estimator of the Extreme Value Parameter. *Extremes*, 1(3), 263-293.

Rootzén, H. and Tajvidi, N. (1997). Extreme Value Statistics and Wind Storm Losses: A Case Study. *Scandinavian Actuarial Journal*, 70-94.

Rosen, O. and Weissman, I. (1996). Comparison of Estimation Methods in Extreme Value Theory. *Communication in Statistics – Theory and Methods*, 25 (4), 759-773.

Scarf, P.A. (1992). Estimation for a Four Parameter Generalized Extreme Value Distribution. *Communication in Statistics – Theory and Method*, 21 (8), 2185-2201.

Segers, J. (2001a). On a Family of Generalized Pickands Estimators. *Preprint*, Katholieke University Leuven.

Segers, J. (2001b). Residual Estimators, *Journal of Statistical Planning and Inference*, 98, 15-27.

Segers, J. and Teugels, J. (2000). Testing the Gumbel Hypothesis by Galton's Ratio. *Extremes*, 3(3), 291-303.

Smith, R.L. (1985). Maximum-Likelihood Estimation in a Class of Non-Regular Cases. *Biometrika*, 72, 67-90.

Smith, R.L. (1989). Extreme Value Analysis of Environmental Time Series: An Application to Trend Detection in Ground-Level Ozone. *Statistical Science*, 4 (4), 367-393.

Yun, S. (2000). A Class of Pickands-Type Estimators for the Extreme Value Index. *Journal of Statistical Planning and Inference*, 83, 113-124.

Received: January 2002, Revised: June 2002

13
ON CONVEX SETS OF MULTIVARIATE DISTRIBUTIONS AND THEIR EXTREME POINTS

C.R. Rao
Department of Statistics
Pennsylvania State University

M. Bhaskara Rao
Department of Statistics
North Dakota State University, Fargo

D.N. Shanbhag
Statistics Division, Department of Mathematical Sciences
University of Sheffield, UK

1. Introduction

Suppose F_1 and F_2 are two n-variate distribution functions. It is clear that any convex combination of F_1 and F_2 is again an n-variate distribution function. More precisely, if $0 \leq \lambda \leq 1$, then the function $\lambda F_1 + (1-\lambda)F_2$ is also a distribution function. A collection $\mathfrak{I}$ of n-variate distributions is said to be convex if for any given distributions F_1 and F_2 in $\mathfrak{I}$, every convex combination of F_1 and F_2 is also a member of $\mathfrak{I}$. An element F in a convex set $\mathfrak{I}$ is said to be an extreme point of $\mathfrak{I}$ if there is no way we can find two distinct distributions F_1 and F_2 in $\mathfrak{I}$ and a number $0 < \lambda < 1$ such that $F = \lambda F_1 + (1-\lambda)F_2$.

Verifying whether or not a collection of multivariate distributions is convex is, usually, fairly easy. For example, the collection of all n-variate normal distributions is not convex. On the other hand, the collection of all n-variate distributions is convex. The difficult task is identifying all the extreme points of a given convex set. However, for the set of all n-variate distributions, any distribution which gives probability mass unity at a single point of the n-dimensional euclidean space is an extreme point and every extreme point of the set is one like this.

The notion of a convex set and its extreme points is very general. Under some reasonable conditions, one can write every point in the convex set as a mixture or a convex combination of its extreme points. Suppose $C = \{(x, y) \in R^2 ; 0 \leq x \leq 1 \text{ and } 0 \leq y \leq 1\}$. The set C is, in fact, the rim and the interior of the unit square in R^2 with vertices (0,0), (0,1), (1,0), and (1,1) and is indeed a convex subset of R^2. The four vertices are precisely the extreme points of the set C. If (x,y) is any given point of C, then we can find four non-negative numbers $\lambda_1, \lambda_2, \lambda_3$, and λ_4 with sum equal to unity such that $(x,y) = \lambda_1(0,0) + \lambda_2(1,0) + \lambda_3(0,1) + \lambda_4(1,1)$. In other words, the element (x,y) in C is a convex combination of the extreme points of C.

In a similar vein, under reasonable conditions, every member of a given convex set $\mathfrak{J}$ of distributions can be written as a mixture or convex combination of the extreme points of $\mathfrak{J}$. Depending on the circumstances, the mixture could take the hue of an integral, i.e., a generalized convex combination! If the number of extreme points is finite, the mixture is always a convex combination. If the number of extreme points is infinite, mixtures can be expressed as an integral.

As an example of integral mixtures, it is well-known that the distributions of exchangeable sequences of random variables constitute a convex set and that the distributions of independent identically distributed sequences of random variables constitute the collection of all its extreme points. Thus one can write the distribution of any sequence of exchangeable sequence of random variables as an integral mixture of distributions of independent identically distributed sequences of random variables. This representation has a bearing on certain characterization problems in distribution theory. Radhakrishna Rao and Shanbhag (1995) explore this phenomenon in their book.

There are some advantages in making an effort to enumerate all extreme points of a given convex set $\mathfrak{J}$ of distributions. Suppose all the extreme points of $\mathfrak{J}$ possess a particular property. Suppose the property is preserved under mixtures. Then we can claim that every distribution in the set $\mathfrak{J}$ possesses the same property. Sometimes, some computations involving a given distribution in the set $\mathfrak{J}$ can be simplified using similar computations for the extreme point distributions. We will see an instance of this phenomenon. The focus of this chapter is in three specific themes of distributions.

- Positive Quadrant Dependent Distributions.
- Distributions which are Totally Positive of Order Two.
- Regression Dependence.

2. Positive Quadrant Dependent Distributions

In this section, we look at two-dimensional distributions. Let X and Y be two random variables with some joint distribution function F. The random variables are said to be positive quadrant dependent or F is positive quadrant dependent if $P(X \geq c, Y \geq d) \geq P(X \geq c)P(Y \geq d)$ for all real numbers c and d. If X and Y are independent, then, obviously, X and Y are positive quadrant dependent. For various properties of positive quadrant dependence, see Lehmann (1966) or Eaton (1982).

13. ON CONVEX SETS OF MULTIVARIATE DISTRIBUTIONS

Let F_1 and F_2 be two fixed univariate distribution functions. Let $M(F_1, F_2)$ be the collection of all bivariate distribution functions F whose first marginal distribution is F_1 and the second marginal distribution is F_2. The set $M(F_1, F_2)$ is a convex set and has been intensively studied in the literature. See Kellerer (1984) and the references therein. See also Rachev and Ruschendorf (1998 a, b). Let $M_{PQD}(F_1,F_2)$ be the collection of all bivariate distributions F whose first marginal distribution is F_1, the second marginal distribution is F_2, and F is positive quadrant dependent. It is not hard to show that the set $M_{PQD}(F_1,F_2)$ is convex. See Bhaskara Rao, Krishnaiah, and Subramanyam (1987). We will now focus on subsets of $M_{PQD}(F_1,F_2)$. Suppose F_1 has support $\{1,2,\ldots,m\}$ with probabilities $p_1, p_2, \ldots, p_m$ and F_2 has support $\{1,2,\ldots,n\}$ with probabilities $q_1, q_2, \ldots, q_n$. The set $M_{PQD}(F_1, F_2)$ can be denoted simply by $M_{PQD}(p_1, p_2, \ldots, p_m, q_1, q_2, \ldots, q_n)$. Every member of $M_{PQD}(p_1, p_2, \ldots, p_m, q_1, q_2, \ldots, q_n)$ can be identified as a matrix $P=(p_{ij})$ of order $m \times n$ such that each entry p_{ij} is non-negative with row sums equal to $p_1, p_2, \ldots, p_m$, column sums equal to $q_1, q_2, \ldots, q_n$, and the joint distribution is positive quadrant dependent. The property of positive quadrant dependence translates into a bunch of inequalities involving p_{ij}'s, p_i's, and q_j's. Let us look at a simple example. Take $m = 2$ and $n = 3$. Let p_1, p_2, q_1, q_2, and q_3 be five positive numbers given satisfying $p_1 + p_2 = 1 = q_1 + q_2 + q_3$. Take any matrix $P=(p_{ij})$ from $M_{PQD}(p_1,p_2,q_1,q_2,q_3)$. The entries of P must satisfy the following two inequalities.

$$p_2 q_3 \leq p_{23} \leq p_2 \wedge q_3 \tag{1}$$

and

$$(p_2 q_2 + p_2 q_3) \vee p_{23} \leq p_{22} + p_{23} \leq p_2 \wedge (q_2 + p_{23}), \tag{2}$$

where $a \vee b$ indicates the maximum of the numbers a and b and $a \wedge b$ indicates the minimum of the numbers a and b. Conversely, let p_{22} and p_{23} be any two non-negative numbers satisfying the above inequalities. Construct the matrix

$$P = \begin{pmatrix} q_1 - p_2 + p_{22} + p_{23} & q_2 - p_{22} & q_3 - p_{23} \\ p_2 - p_{22} - p_{23} & p_{22} & p_{23} \end{pmatrix}$$

of order 2×3. One can show that $P \in M_{PQD}(p_1,p_2,q_1,q_2,q_3)$. The implication of this observation is that the numbers p_{22} and p_{23} determine whether or not the matrix P constructed above belongs to $M_{PQD}(p_1,p_2,q_1,q_2,q_3)$. The two inequalities (1) and (2) determine a simplex in the $p_{22} - p_{23}$ plane. Let us look at a very specific example: $p_1 = p_2 = \frac{1}{2}$ and $q_1 = q_2 = q_3 = 1/3$. The inequalities (1) and (2) become

$$1/6 \leq p_{23} \leq 1/3$$

and

$$1/3 = p_{23} \vee 1/3 \leq p_{22} + p_{23} \leq (1/2) \wedge (1/3 + p_{23}) = \frac{1}{2}.$$

There are four extreme points of the set $M_{PQD}(1/2, \frac{1}{2}, 1/3, 1/3, 1/3)$ given by

$$\begin{pmatrix} 1/6 & 1/6 & 1/6 \\ 1/6 & 1/6 & 1/6 \end{pmatrix}, \begin{pmatrix} 1/6 & 1/3 & 0 \\ 1/6 & 0 & 1/3 \end{pmatrix}, \begin{pmatrix} 1/3 & 1/6 & 0 \\ 0 & 1/6 & 1/3 \end{pmatrix}, \text{ and}$$
$$\begin{pmatrix} 1/3 & 0 & 1/3 \\ 0 & 1/3 & 1/6 \end{pmatrix}.$$

Every member P of $M_{PQD}(1/2,1/2,1/3,1/3,1/3)$ is a convex combination of these four matrices. Even for moderate values of m and n, determining the extreme points of $M_{PQD}(p_1,p_2,\ldots,p_m, q_1,q_2,\ldots,q_n)$ could become very laborious. A method of determining the extreme points of $M_{PQD}(p_1,p_2,\ldots,p_m, q_1,q_2,\ldots,q_n)$ has been outlined in Bhaskara Rao, Krishnaiah, and Subramanyam (1987).

3. Distributions which are Totally Positive of Order Two

Let X and Y be two random variables each taking a finite number of values. For simplicity, assume that X takes values $1,2,\ldots,m$ and Y takes values $1,2,\ldots,n$. Let $p_{ij} = \Pr(X=i, Y=j)$, $i = 1,2,\ldots,m$ and $j = 1,2,\ldots,n$. Let $P = (p_{ij})$ be the matrix of order $m \times n$, which gives the joint distribution of X and Y.

The random variables X and Y are said to be totally positive of order 2 (TP_2) (or, equivalently, P is said to be totally positive of order two) if the determinants

$$\begin{vmatrix} p_{i_1j_1} & p_{i_1j_2} \\ p_{i_2j_1} & p_{i_2j_2} \end{vmatrix} \geq 0$$

for all $1 \leq i_1 < i_2 \leq m$ and $1 \leq j_1 < j_2 \leq n$. In the literature, this notion also goes by the name positive likelihood ratio dependence. See Lehmann (1966). For some ramifications of this definition, see Barlow and Proschan (1981).

We now assume that $m = 2$. Let $q_1,q_2,\ldots,q_n$ be n given positive numbers with sum equal to unity. Let $q = (q_1,q_2,\ldots,q_n)$. Let $M_q(TP_2)$ be the collection of all matrices P of order $2 \times n$ with column sums equal to $q_1,q_2,\ldots,q_n$ and P is totally positive of order two. It is not hard to show that $M_q(TP_2)$ is convex. See Subramanyam and Bhaskara Rao (1988). The extreme points of these convex sets have been identified precisely. In fact, the convex set $M_q(TP_2)$ has $n+1$ extreme points and are given by $P_0 = \begin{pmatrix} 0 & 0 & & 0 \\ q_1 & q_2 & \cdots & q_n \end{pmatrix}$ and

$$P_i = \begin{pmatrix} q_1 & q_2 & \cdots & q_i & 0 & 0 & \cdots & 0 \\ 0 & 0 & \cdots & 0 & q_{i+1} & q_{i+2} & \cdots & q_n \end{pmatrix}, \; i = 1,2,\ldots,n.$$ The determination of extreme points in the general case of $m \times n$ matrices is still an open problem.

4. Regression Dependence

Let X and Y be two random variables. Assume that X takes values $1,2,\ldots,m$ and Y takes values $1,2,\ldots,n$. Let $p_{ij} = \Pr(X=i, Y=j)$, $i = 1,2,\ldots,m$ and $j =$

1,2,...,n. Let $P = (p_{ij})$. Say that Y is strongly positive regression dependent on $X(\text{SPRD}(Y|X))$ if $\text{PR}(Y \le j|X = i)$ is non-increasing in i for each j. See Lehmann (1966) and Barlow and Proschan (1981) for an exposition of this notion. This notion also goes by the name that Y is stochastically increasing in X. We now assume that $m = 2$. In this case, Y is strongly positive regression dependent on X is equivalent to the notion that X and Y are positive quadrant dependent. Let p_1 and p_2 be two given positive numbers with sum equal to unity. Let $M_{PQD}(n,p_1,p_2)$ be the collection of all matrices $P = (p_{ij})$ of order $2 \times n$ such that the row sums are equal to p_1 and p_2 and P is positive quadrant dependent. Equivalently, with respect to the joint distribution P, Y is strongly positive regression dependent on X. We would like to report some new results on the set $M_{PQD}(n,p_1,p_2)$.

<u>Theorem 1</u>: The set $M_{PQD}(n,p_1,p_2)$ is convex.

<u>Theorem 2</u>: The total number of extreme points of $M_{PQD}(n,p_1,p_2)$ is $n(n+1)/2$ and these are given by

$$\begin{pmatrix} p_1 & 0 & \cdots & 0 \\ p_2 & 0 & \cdots & 0 \end{pmatrix}, \begin{pmatrix} p_1 & 0 & 0 & \cdots & 0 \\ 0 & p_2 & 0 & \cdots & 0 \end{pmatrix}, \ldots, \begin{pmatrix} p_1 & 0 & \cdots & 0 \\ 0 & 0 & \cdots & p_2 \end{pmatrix},$$

$$\begin{pmatrix} 0 & p_1 & 0 & \cdots & 0 \\ 0 & p_2 & 0 & \cdots & 0 \end{pmatrix}, \ldots, \begin{pmatrix} 0 & p_1 & & 0 \\ 0 & 0 & \cdots & p_2 \end{pmatrix},$$

$$\cdots \quad \cdots \quad \cdots \quad \cdots$$

$$\begin{pmatrix} 0 & 0 & \cdots & p_1 \\ 0 & 0 & \cdots & p_2 \end{pmatrix}.$$

Theorem 1 is not hard to establish. The proof of Theorem 2 is rather involved. The details will appear elsewhere.

5. An Application

Suppose X and Y are discrete random variables with unknown joint distribution. Suppose we wish to test the null hypothesis H_0: X and Y are independent against the alternative H_1: X and Y are positive quadrant dependent but not independent. Suppose the marginal distributions of X and Y are known. Let $(X_1,Y_1), (X_2,Y_2),\ldots,(X_N,Y_N)$ be N independent realizations of (X,Y). Let T be a statistic (a function of the given data) that is used to test the validity of H_0 against H_1. Let C be the critical region of the test. We want to compute the power function of the test based on T. Let μ be a specific joint distribution of X and Y under H_1. The computation of $\Pr(T \in C)$ under the joint distribution μ of X and Y is usually difficult. Let $\mu_{(1)}, \mu_{(2)},\ldots,\mu_{(k)}$ be the extreme points of the set of all distributions of X and Y with the given marginals which are positive quadrant dependent. One can write $\mu = \lambda_1 \bullet \mu_{(1)} + \lambda_2 \bullet \mu_{(2)} + ... + \lambda_k \bullet \mu_{(k)}$ for some non-negative λ's with sum equal to unity. The joint distribution of the data is given by the product measure μ^N. One can write the joint distribution of the data as a convex combination of joint distributions of the type $\mu_{(1)}^{n_1} \otimes \mu_{(2)}^{n_2} \otimes ... \otimes \mu_{(k)}^{n_k}$ with $n_1, n_2, ..., n_k$ non-negative integers with sum equal to N. Generally, computing $\Pr(T \in C)$ is easy under such special joint distributions. Also, $\Pr(T \in C)$ under μ^N will be a convex combination of $\Pr(T \in C)$'s evaluated under such special joint distributions. The relevant formula has been hammered out in Bhaskara Rao, Krishnaiah, and Subramanyam (1987).

6. References

Barlow, R.E. and Proschan, F. (1981). Statistical Theory of Reliability and Life Testing: Probability Models. Holt, Reinhart, and Winston, Silver Spring, Maryland.
Eaton, M.L. (1982). A Review of Selected Topics in Multivariate Probability Inequalities, *Annals of Statistics*, 10, 11-43.
Kellerer, H.G. (1984). Duality Theorems for Marginal Problems, Z. Wahrscheinkeitstheorie Verw. Geb., 67, 399-432.
Lehmann, E.L. (1966). Some Concepts of Dependence, *Ann. Math. Statist.*, 37, 1137-1153.
Rachev, S.T. and Ruschendorf, L. (1998a). Mass Transportation Problems, Part I : Theory. Spinger-Verlag, New York.
Rachev, S.T. and Ruschendorf, L. (1998b). Mass Transportation Problems, Part II : Applications. Spinger-Verlag, New York.
Rao, M. Bhaskara, Krishnaiah, P.R., and Subramanyam, K. (1987). "A Structure Theorem on Bivariate Positive Quadrant Dependent Distributions and Tests for Independence in Two-Way Contingency Tables." *J. Multivariate Anal.*, 23, 93-118.
Rao, C. Radhakrishna and Shanbhag, D.N. (1995). Choquet-Deny Type Functional Equations with Applications to Stochastic Models. John Wiley, New York.
Subramanyam, K. and Rao, M. Bhaskara (1988). Analysis of Odds Ratios in 2×n Ordinal Contingency Tables. *J. Multivariate Anal.*, 27, 478-493.

Received: September 2001

14
THE LIFESPAN OF A RENEWAL

Jef L. Teugels
Department of Mathematics
Katholieke Universiteit Leuven, Heverlee, Belgium

1. Introduction

We start with a given renewal process in which we define the quantities that will play a role in what follows. The concepts can be found in a variety of textbooks like Alsmeyer (1991), Feller (1971), Karlin and Taylor (1975), and Ross (1983).

Definition 1. Let $\{X_i; i \in \mathbf{N}\}$ be a sequence of independent identically distributed random variables with common distribution F of X, where $X \geq 0$ but $X \not\equiv 0$. The sequence $\{X_i; i \in \mathbf{N}\}$ is called a *RENEWAL PROCESS*. Let $S_0 = 0$ and $S_n = X_n + S_{n-1}$, $(n \geq 1)$. Then the sequence $\{S_n; n \in \mathbf{N}\}$ constitutes the set of *RENEWAL (TIME) POINTS*. Let also $t \geq 0$, and define $N(t)=\sup[n: S_n \leq t]$; then the process $\{N(t); t \geq 0\}$ is called the *RENEWAL COUNTING PROCESS*. The *RENEWAL FUNCTIONS* are defined and denoted by

$$U(t) =: EN(t) = \sum_{k=1}^{\infty} F^{*k}(t), \quad U_o(t) =: I(t) + U(t) = \sum_{k=1}^{\infty} F^{*k}(t).$$

We say that U and/or U_o are renewal functions *GENERATED* by F or by X.
We recall from general renewal theory that the renewal functions satisfy the renewal equations $U = F + F * U$ and $U_o = I + F * U_o$.

Definition 2. Let $\{S_i, i \in \mathbf{N}\}$ and $\{N(t), t \geq 0\}$ be as defined above. For every $t \geq 0$ we define
(i) the *AGE* of the renewal process by $Z(t) = t - S_{N(t)}$;
(ii) the *RESIDUAL LIFE (TIME)* of the renewal process by $Y(t)=S_{N(t)+1}-t$;
(iii) the *LIFESPAN* of the renewal process by $L(t) = Y(t) + Z(t) = X_{N(t)+1}$.

The most important theoretical result in renewal theory is *Blackwell's theorem* which is often phrased in its more general key *renewal theorem* form. To get the best formulation, we follow Feller (1971). Let z be a real-valued function on $\mathfrak{R}_+$. Let h be any positive real number. Define for $n \geq 0$

$$\overline{m}_n = \sup\{z(x) : nh \leq x \leq (n+1)h\},$$

$$\underline{m}_n = \inf\{z(x) : nh \leq x \leq (n+1)h\}.$$

We call z *directly Riemann-integrable* if

(i) (a) $\sum_{n=0}^{\infty} |\overline{m}_n| < \infty$, (b) $\sum_{n=0}^{\infty} |\underline{m}_n| < \infty$;

(ii) $\limsup_{h \downarrow 0} h \sum_{n=0}^{\infty} \{\overline{m}_n - \underline{m}_n\} = 0.$

Theorem 1. *The Key Renewal Theorem.* Assume that F generates a renewal process and that $\mu := EX < \infty$. Let m be directly Riemann-integrable. Then if F is non-lattice, as $x \uparrow \infty$,

$$m * U(x) \to \frac{1}{\mu} \int_0^{\infty} m(x)dx \quad .$$

In particular, if m is non-negative, non-increasing and improper Riemann-integrable, then m is also directly Riemann-integrable.

2. General Properties

We begin by giving the distribution and the mean of L(t). See for example Ross (1983).

Theorem 2. The lifespan of a renewal process satisfies the following properties.

(i) For $t \geq 0$ and $y \geq 0$ we have

$$P\{L(t) \leq y\} = \begin{cases} 1 - \{1 - F(y)\}U_0(t) & \text{if } t \leq y, \\ \int_{t-y}^{t} \{F(y) - F(t-x)\}dU(x) & \text{if } y \leq t. \end{cases}$$

(ii) If $\mu < \infty$, then

$$\lim_{t \uparrow \infty} P\{L(t) \leq y\} = \frac{1}{\mu} \int_0^y u dF(u) =: \tilde{F}(y).$$

(iii) EL(t) satisfies the renewal equation

$$EL(t) = \int_t^{\infty} x dF(x) + EL * F(t),$$

if $\text{Var}(X) < \infty$, then

14. THE LIFESPAN OF A RENEWAL

$$\lim_{t\uparrow\infty} EL(t) = \mu + \frac{Var(X)}{\mu}.$$

Proof:
(i) The proof of the first part is easily accomplished by a renewal argument. For

$$P\{L(t) > y\} = \int_0^\infty P\{L(t) > y \mid S_1 = x\} dF(x).$$

However,

$$P\{L(t) > y \mid S_1 = x\} = \begin{cases} 0 & \text{if } t < x \le y, \\ 1 & \text{if } t < x \text{ and } y < x, \\ P\{L(t-x) > y\} & \text{if } x \le t. \end{cases}$$

Indeed, if $S_1 = x > t$, then $L(t) = S_1$ which explains the first two alternatives. The remaining case is obtained by shifting time to the new point x. The resulting equation reads as follows

$$P\{L(t) > y\} = \{1 - F(y \vee t)\} + \int_0^t P\{L(t-x) > y\} dF(x).$$

The solution to this equation yields the promised result.

(ii) If we compare (i) with the statement of the key renewal theorem, it is natural to identify $m(u) = \{F(y) - F(u)\} I_{[0,y]}(u)$; then $P\{L(t) \le y\} = m*U_o(t)$ and the results follows from Th 1.

(iii) This formula can easily be obtained by integration of (i) and by recalling that $EL(t) = \int_0^\infty P\{L(t) > x\} dx.$

Theorem 3. If $\beta \ge 0$, then for $t \uparrow \infty$

$$EL^\beta(t) \to \frac{1}{\mu} E(X^{\beta+1})$$

whenever the right-hand side is finite.

Proof: It is well known that for a non-negative random variable W, and $\beta > 0$,

$$EW^\beta = \beta \int_0^\infty x^{\beta-1} P(W > x) dx.$$

From the equation in the middle of the preceding proof one can then easily derive that with $E^\beta(t) =: V_\beta(t)$

$$V_\beta(t) = g_\beta(t) + V_\beta * F(t)$$

where in turn

$$g_\beta(t) = \beta \int_0^\infty y^{\beta-1}(1 - F(y \vee t))dy.$$

Solving the latter equation for $V_\beta(t)$ gives $V_\beta(t) = g_\beta * U_o(t)$ with $U_o(t)$ the renewal function. From the key renewal with $m = g_\beta$ it follows that we only have to evaluate the integral

$$\int_0^\infty g_\beta(t)dt = \beta \int_0^\infty dt \int_0^t y^{\beta-1}(1 - F(t))dy + \beta \int_0^\infty dt \int_t^\infty y^{\beta-1}(1 - F(y))dy$$

$$= \int_0^\infty t^\beta (1 - F(t))dt + \beta \int_0^\infty y^{\beta-1}(1 - F(y))dy \int_0^y dt$$

$$= (\beta + 1) \int_0^\infty y^\beta (1 - F(y))dy$$

$$= EX^{\beta+1}.$$

The case $\beta = 1$ has been covered by Th 2. For the variance one needs a little bit of calculation to show that if $EX^3 < \infty$, then

$$\text{VarL}(t) \to \frac{1}{\mu}E(X - \mu)^3 - \left(\frac{\text{Var}(X)}{\mu}\right)^2 (\text{Var}X - \mu^2).$$

In a similar fashion as in Th 2 one can actually prove a result that involves all three of the random variables $Y(t)$, $Z(t)$ and $L(t)$. We quote from Hinderer (1985).

Lemma 1. Let G be any Borelset in $\Re_+ \times \Re_+ \times \Re_+$. Then for $t \geq 0$ we have

$$P\{(Y(t), Z(t), L(t)) \in G\} = \int_0^t P\{(X + u - t, t - u, X) \in G, X > t - u\}dU_o(u).$$

In particular, $P\{Z(t) \leq z\} = \int_{t-z}^t \{1 - F(t - u)\}dU_o(u)$ for $0 \leq z \leq t$.

Example: *The exponential case.* If the renewal process is generated by an exponential distribution with density $f(x) = \lambda \exp(-\lambda x)$ on $\Re_+$, then an easy calculation yields

$$P\{L(t) \leq y\} = 1 - e^{-\lambda y}\{1 + \lambda(t \vee y)\}.$$

This means that in the exponential case L(t) has a discontinuity at the point t since

$$f_{L(t)}(y) = \begin{cases} \lambda e^{-\lambda y}(1+\lambda t) & \text{if } t < y, \\ \lambda^2 y e^{-\lambda y} & \text{if } y \leq t. \end{cases}$$

This result leads immediately to the expression

$$EL(t) = \frac{1}{\lambda}\{2 - (1+\lambda t)e^{-\lambda t}\}.$$

The latter expression shows that $EL(0) = EX = \lambda^{-1}$, but as time develops, $EL(t)$ increases gradually to $EL(\infty) = 2EX$. This result illustrates already one possible form of the *renewal paradox* that we discuss in more detail in the next section.

3. The Renewal Paradox

In this section we show why $L(t) = X_{N(t)+1}$ is a lifetime that is very different from the lifetimes that make up the renewal process. We do this by gradually sharpening the information about the difference between these two quantities. Again, we basically follow Ross (1983).

(1) We first note that (iii) of Th 2 above implies that $\lim_{t \uparrow \infty} L(t) = L(\infty) \geq EX = \mu$. The difference between the two sides depends on the variance of the generic distribution. Loosely speaking, if this variance is large, it seems likely that the renewal that is ultimately at work is one with a large expectation.

(2) We quantify the difference further. To do that we prove that

$$\forall\, t \geq 0 : EL(t) \geq \mu.$$

As an intermediate step let us prove that for all $t \geq 0$, $\widetilde{F}(t) \leq F(t)$, a result interesting in its own right. For if $t \leq \mu$ then $\mu \widetilde{F}(t) = \int_0^t u\, dF(u) \leq tF(t) \leq \mu F(t)$. If $t \geq \mu$, then $\mu(1-\widetilde{F}(t)) = \int_t^{\infty} u\, dF(u) \geq t(1-F(t)) \geq \mu(1-F(t))$ yielding again the required inequality.

If we solve the renewal equation in (iii) of Th 2 above, then it easily follows that

$$\frac{EL(t)}{\mu} - 1 = U(t) - \widetilde{F}(t) - \widetilde{F} * U(t).$$

By the monotonicity of U we find that the right-hand side is underestimated by U(t)-F(t)-F*U(t) which is actually 0 by the renewal equation for U.

(3) We strengthen the inequality even further by proving
$$\forall\, t \geq 0,\, \forall\, y \geq 0 : P\{L(t) > y\} \geq P\{X > y\}.$$

One way of expressing the content of the latter inequality is by saying that $L(t)$ is *stochastically larger* than X or that X is *stochastically dominated* by $L(t)$. For this type of ordering among random variables one often writes that $X \prec L(t)$.

To prove the inequality take the result (i) in Th 2. If $t \leq y$ then since $U_o(t) \geq 1$, $P\{(L(t) > y\} \geq \{1\text{-}F(t)\}$. When $y < t$, write

$$P\{L(t) \leq y\} = \int_{t-y}^{t}\{1 - F(t-x)\}dU(x) - \int_{t-y}^{y}\{1 - F(y)\}dU(x)$$

$$= P\{Z(t) \leq y\} - \{1 - F(y)\}\{U(t) - U(t-y)\}$$

by using lemma 1. From that same formula we note that $P\{Z(t) \leq y\} \leq U(t) - U(t-y)$ so that

$$P\{L(t) \leq y\} \leq P\{Z(t) \leq y\}\{1\text{-}(1 - F(y))\}.$$

It is obvious that the third form implies the second as is shown by taking expectations. The fact that $L(t)$ has a fatter tail than the generic variable is generally called the *renewal paradox*. Considering the renewal at work at t as a regular copy of the generic variable leads to positive biases and is therefore wrong. In the case where $EX < \infty$ but $Var(X) = \infty$, $EL(t)$ even tends to ∞.

4. The Infinite Mean Case

If the mean $\mu = \infty$, it is classical to assume that the generic distribution is of Pareto-type, i.e., for some slowly varying function $\ell(x)$ and an index of regular variation $\alpha \in (0, 1)$

$$1\text{-}F(x) \sim x^{-\alpha}\ell(x)$$

for $x \uparrow \infty$. From general renewal theory we recall the *elementary renewal theorem for the null-recurrent case*. This result has been originally given in Dynkin (1961) and in Lamperti (1962). For a treatment and a proof, we refer to Bingham, et al. (1987).

<u>Theorem 4.</u> Let $\ell(x)$ be slowly varying and $0 \leq \alpha < 1$. Then as $x \uparrow \infty$ the following conditions are equivalent:
(i) $1\text{-}F(x) \sim x^{-\alpha}\ell(x)$,
(ii) $U(x) \sim \dfrac{\sin(\alpha\pi)}{\alpha\pi} x^{\alpha}\ell^{-1}(x)$.

They both imply that

(iii) $\lim_{x \uparrow z} \{1 - F(x)\} U(x) = \dfrac{\sin(\alpha\pi)}{\alpha\pi}$.

The above theorem has an immediate consequence, known as the *Dynkin-Lamperti theorem*.

Theorem 5. Assume $1-F(x) \sim x^{-\alpha} \ell(x)$ where $0<\alpha<1$ and ℓ slowly varying.

(i) $\left(\dfrac{Y(t)}{t}, \dfrac{Z(t)}{t} \right) \xrightarrow{D} (Y, Z)$ as $t \uparrow \infty$, where (Y, Z) have a joint density on $\mathfrak{R}_+ \times (0, 1)$ given by

$$p_\alpha(y,z) = \dfrac{\alpha \sin(\alpha\pi)}{\pi} (1-z)^{\alpha-1} (y+z)^{-(\alpha+1)};$$

(ii) Also $L(t)/t$ has a limiting density given by

$$r_\alpha(x) = \dfrac{\sin(\alpha\pi)}{\pi} x^{-(\alpha+1)} g(x)$$

where in turn

$$g_\alpha(x) = \begin{cases} 0 & \text{if } x \leq 0, \\ 1 - (1-x)^\alpha & \text{if } 0 \leq x \leq 1, \\ 1 & \text{if } 1 \leq x \end{cases}.$$

Proof:
(i) The proof is classical. Take $0 < y$ and $0 < z < 1$. Start from lemma 1 and take $G=(0, yt) \times (0, zt) \times \mathfrak{R}_+$. Then

$$P\{Y(t) \leq yt, Z(t) \leq zt\} = \int_{t-z}^{t} P\{t - u < X \leq ty + t - u\} dU_o(u)$$

$$= \int_{1-z}^{1} \dfrac{F(t(y+1-u)) - F(t(1-u))}{1 - F(t)} (1 - F(t)) U_o(t) d\dfrac{U_o(ut)}{U_o(t)}$$

$$\to \dfrac{\sin\alpha\pi}{\pi} \int_{1-z}^{1} \{(1-u)^{-\alpha} - (1-u+y)^{-\alpha}\} u^{\alpha-1} du$$

where the limit has to be taken with the necessary care using all three statements of Th 3. For more details consult Bingham et al. (1987).
For the given region, the right-hand side is the joint distribution of the bivariate density given in the formulation of the theorem.

(ii) This result can be obtained by integrating the joint density in part (i). Alternatively, use (i) of Th 2. First, with $y=tz$ and for $z \geq 1$.

$$P\{L(t) \geq tz\} = \frac{1-F(zt)}{1-F(t)}(1-F(t))U_o(t) \to z^{-\alpha}\frac{\sin\alpha\pi}{\alpha\pi}$$

which has the density as given in the statement. Next, for 0<z<1,

$P\{L(t) \leq zt\} =$

$$= \int_{1-z}^{1}\frac{F(zt)-F(t(1-u))}{1-F(t)}(1-F(t))U_o(t)d\frac{U_o(tu)}{U_o(t)}$$

$$\to \frac{\sin\alpha\pi}{\alpha\pi}\int_{1-z}^{1}\{(1-u)^{-\alpha}-z^{-\alpha}\}du^{\alpha}.$$

Taking derivatives with respect to z on the right-hand side proves the remaining case.

5. Normalized Age and Residual Life

Since $Y(t) + Z(t) = L(t)$, the lifespan can be used to normalize the age and the residual life of a renewal process. In the same vain as in Lemma 1 one can prove the following.

Lemma 2. Let $0 \leq y \leq 1$, $0 < z \leq 1$ and $1 < y+z$. Then

$$P\left\{\frac{Y(t)}{L(t)} \leq y, \frac{Z(t)}{L(t)} \leq z\right\} = \int_{0}^{t}\left\{F\left(\frac{t-u}{1-y}\right) - F\left(\frac{t-u}{z}\right)\right\}dU_o(u).$$

This result can be used to derive the asymptotic behavior of the joint distribution.

Theorem 6. Let $t \uparrow \infty$ and $0 < y, z < 1$, $1 < y+z$.
(i) If $\mu < \infty$, then

$$P\left\{\frac{Y(t)}{L(t)} \leq y, \frac{Z(t)}{L(t)} \leq z\right\} \to y+z-1.$$

(ii) If $1-F(x) \sim x^{-\alpha}\ell(x)$ with $0 \leq \alpha < 1$ and ℓ slowly varying, then

$$P\left\{\frac{Y(t)}{L(t)} \leq y, \frac{Z(t)}{L(t)} \leq z\right\} \to z^{\alpha} - (1-y)^{\alpha}.$$

Proof:

(i) This follows immediately from the key renewal theorem by the choice

$$m(x) = F\left(\frac{x}{1-y}\right) - F\left(\frac{x}{z}\right).$$

(ii) As is the previous proofs, write

$$P\left\{\frac{Y(t)}{L(t)} \le y, \frac{Z(t)}{L(t)} \le z\right\} = \int_0^1 \frac{F\left(\frac{t(1-w)}{1-y}\right) - F\left(\frac{t(1-w)}{z}\right)}{1-F(t)}(1-F(t))U_o(t)d\frac{U_o(tw)}{U_o(t)}$$

$$\to \frac{\sin\alpha\pi}{\alpha\pi}\int_0^1\left[\left(\frac{1-w}{z}\right)^{-\alpha} - \left(\frac{1-w}{1-y}\right)^{-\alpha}\right]dw^\alpha$$

$$= z^\alpha - (1-y)^\alpha.$$

Note that in the first case, the two marginals are uniform. Also, for the boundary case $\alpha=1$, the two results coincide.

6. An Application

In utility theory one often encounters the quantity $E(X\varphi(X))$ for some normalized function φ. Such functions are solutions of extremal problems. See for example Goovaerts et al. (1984). The following problem is therefore of some independent interest. Suppose one observes a Poisson process generated by an exponential distribution with arbitrary parameter λ. Determine the *random time* T with distribution G such that

$$EL(T) = \frac{E(XG(X))}{EG(X)}.$$

If we refer to example 1, we notice that the left-hand side can be expressed in terms of $\hat{G}(s)$, the Laplace transform of T. For,

$$EL(T) = E\left(\frac{1}{\lambda}\{2 - (1-\lambda T)e^{-\lambda T}\}\right)$$

$$= \frac{2}{\lambda} - \frac{1}{\lambda}\hat{G}(\lambda) + \hat{G}'(\lambda).$$

But also the right-hand side can be written in terms of $\hat{G}$. Indeed,

$$EG(X) = \int_0^\infty G(x)d(1-e^{-\lambda x}) = \hat{G}(\lambda)$$

as follows immediately by an integration by parts. Similarly,

$$E(XG(X)) = \frac{1}{\lambda}\hat{G}(\lambda) + \hat{G}'(\lambda).$$

As λ is arbitrary, we need to solve the first order differential equation

$$\hat{G}(s) + \hat{G}^2(s) + s\hat{G}'(s)(1+\hat{G}(s)) = 0.$$

Separation of the variables easily leads to the general solutions

$$\hat{G}(s) = 1 + 2\alpha s \pm 2\sqrt{\alpha s(1+\alpha s)}.$$

where α is any non-negative constant. If we request that $\hat{G}(s)$ should be a (completely) monotonic function, then we need to take the negative sign. Consulting a table of Laplace transforms in Abramovitz and Stegun (1965), we see that for b>0

$$\int_0^\infty e^{-st}\{e^{-bt}I_1(bt)\}\frac{dt}{t} = \frac{\sqrt{s+2b}-\sqrt{s}}{\sqrt{s+2b}+\sqrt{s}}$$

where I_1 is a special case of the modified Bessel function

$$I_v(z) = \left(\frac{z}{2}\right)^n \sum_{k=0}^{\infty} \frac{\left(\frac{z^2}{4}\right)^k}{k!\Gamma(v+k+1)}.$$

We therefore find that the requested G has a density of the form

$$\frac{dG(u)}{du} = \frac{1}{u}c^{-\frac{u}{2\alpha}}I_1\left(\frac{u}{2\alpha}\right).$$

7. Some Further Results

We finish with a few statements that illustrate the versatility of the lifespan as a random variable:

(1) An interesting question is whether one can recover anything at all about the renewal process by observing L(t) of even EL(t). Let us assume that $\mu<\infty$. Look at the renewal equation in (iii) of Th 2. Since EL(0)=μ, it seems more advantageous to work with the quantity
$$M(t) =: EL(t) - \mu$$

14. THE LIFESPAN OF A RENEWAL

which is positive for all $t \geq 0$. The renewal equation for $M(t)$ is easily derived, i.e.,

$$M(t) = \mu\{F(t) - \tilde{F}(t)\} + M * F(t).$$

Taking Laplace transforms of both sides we get

$$\hat{M}(s) = \mu\hat{F}(s) + \hat{F}'(s) + \hat{M}(s)\hat{F}(s).$$

This is essentially a non-homogeneous first-order differential equation in $\hat{F}(s)$, i.e.,

$$\hat{F}'(s) + (\mu + \hat{M}(s))\hat{F}(s) = \hat{M}(s).$$

with initial value $\hat{F}(0) = 1$. The solution to this equation is given by

$$\hat{F}(s) = e^{-m(s)}\left\{1 + \int_0^s e^{m(t)}\hat{M}(t)dt\right\}$$

where in turn $m(s) =: \int_0^s \{\mu + \hat{M}(u)\}du.$

Since $\hat{F}$ fully determines F, one can theoretically derive F from a knowledge of $EL(t)$.

(2) A useful but slightly cumbersome quantity is the residual lifespan of the renewal at work. Following the usual nomenclature, one defines the residual lifespan by

$$R_t(x, y) = P\{L(t) \leq x+y \mid L(t) > x\}.$$

Following the same calculations as in the proof of lemma 1, we can show that

$$R_t(x,y) = \begin{cases} (J_t(x))^{-1} \int_x^{x+y} \{U(t) - U(t-u)\}dF(u) & \text{if } 0 \leq x \leq x+y < t, \\ (J_t(x))^{-1} \int_x^{x+y} \{U(t) - U(t-u)_+\}dF(u) & \text{if } x < t \leq x+y, \\ \dfrac{F(x+y) - F(x)}{1 - F(x)} & \text{if } t \leq x < x+y. \end{cases}$$

Here $J_t(x) = P\{L(t) > x\}$ can be obtained from lemma1. Note that for the last case, the right-hand side is only depending on t in the boundary.

When $\mu < \infty$ and $t \uparrow \infty$, then the top formula applies. Blackwell's theorem easily yields that

$$\lim_{t\uparrow\infty} R_t(x,y) = \frac{\widetilde{F}(x+y)-\widetilde{F}(x)}{1-\widetilde{F}(x)}.$$

When however, $\mu = \infty$ and we assume that $1-F(x) \sim x^{-\alpha}\ell(x)$ where $0\leq\alpha<1$, then we need to look at $R_t(tx, ty)$ in order to get a limit. Using the function g_α defined in Th 5 we find that

$$\lim_{t\uparrow\infty} R_t(tx,ty) = 1 - \left(1+\frac{y}{x}\right)^{-\alpha-1} \frac{g_\alpha(x+y)}{g_\alpha(x)}.$$

8. References

Abramovitz, M. and Stegun, I.A. (1965). *Handbook of Mathematical Functions with Formulas, Graphs and Mathematical Tables.* Dover Publications, New York.

Alsmeyer, G. (1991). *Erneuerungstheorie.* B.G. Teubner, Stuttgart.

Bingham, N.H., Goldie, C.M. and Teugels, J.L. (1987). *Regular Variation, Encyclopedia of Mathematics and its Applications.* Cambridge University Press, Cambridge.

Dynkin, E.B. (1961). Some Limit Theorems for Sums of Independent Random Variables with Infinite Mathematical Expectations. *Selected Translations Math. Stat. Prob.*, 1, 171-189.

Feller, W. (1971). *An Introduction to Probability Theory and its Applications,* Vol. 2, 2nd ed. Wiley, New York.

Goovaerts, M.J., De Vylder, F. and Haezendonck, J. (1984). *Insurance Premiums.* North Holland, Amsterdam.

Hinderer, K. (1985). A Unifying Method for Some Computations in Renewal Theory. *Z. Angew. Math. Mech.*, 65, 199-206.

Karlin, S. and Taylor, H.M. (1975). *A First Course in Stochastic Processes,* 2nd ed. Academic Press, New York.

Lamperti, J. (1962). An Invariance Principle in Renewal Theory. *Ann. Math. Statist.*, 33, 685-696.

Ross, S.M. (1983). *Stochastic Processes.* Wiley, New York.

Received: February 2000

15
MAXIMUM LIKELIHOOD ESTIMATES OF GENETIC EFFECTS

Wolfgang Urfer and Katharina Emrich
Department of Statistics
University of Dortmund, Germany

1. Introduction

The investigation of genes affecting economically important agronomic traits has a long tradition in statistics. Along with rapid developments in molecular marker technologies, biometrical models have been constructed, refined, and generalized for detecting, mapping, and estimating the effects of loci that control quantitative traits. Such loci and their genotypes are called 'Quantitative Trait Loci' (QTLs).

A wide range of models and methods have been developed for the investigation of effects of trait loci. Jansen (1996) explains a Monte-Carlo expectation-maximization-algorithm for fitting multiple QTLs to incomplete genetic data. Stephens and Fisch (1998) employ reversible jump Markov-chain Monte-Carlo-methodology to compute posterior densities for the parameters and the number of QTLs. Kao and Zeng (1997) presented Maximum Likelihood estimates using an ECM (Expectation/Conditional Maximization)-algorithm and calculated asymptotic variance-covariance matrices for the estimates.

We consider experimental populations derived from a cross between two parental inbred lines P_1 and P_2, differing mainly in a quantitative trait of interest. This allows for investigating the realizations of alleles of marker loci of a marker interval that contains a putative QTL. Two flanking markers for an interval, where a putative QTL (with alleles Q_1, Q_2) is being tested, have alleles A_1, A_2 and B_1, B_2. If the F_1 individuals are selfed or intermated, an F_2 – population with nine observable marker genotypes is produced.

The realizations of the alleles of the QTLs are not observable. A straightforward calculation of conditional probabilities of the QTL genotypes given the marker genotypes is shown in the following table for the nine different types of marker combinations.

Table 15.1 Conditional probabilities of QTL genotypes given marker genotypes
(Kao & Zeng, 1997)

No.	Marker genotypes	Expected frequencies	QTL - genotypes Q_1/Q_1	Q_1/Q_2	Q_2/Q_2	Sample size n
1	$A_1\ B_1 / A_1\ B_1$	$(1-r)^2/4$	1	0	0	n_1
2	$A_1\ B_1 / A_1\ B_2$	$r(1-r)/2$	$1-p$	p	0	n_2
3	$A_1\ B_2 / A_1\ B_2$	$r^2/4$	$(1-p)^2$	$2p(1-p)$	p^2	n_3
4	$A_2\ B_1 / A_1\ B_1$	$r(1-r)/2$	p	$1-p$	0	n_4
5	$A_2\ B_2 / A_1\ B_1$	$(1-r)^2/2 + r^2/2$	$cp(1-p)$	$1-2cp(1-p)$	$cp(1-p)$	n_5
6	$A_2\ B_2 / A_1\ B_2$	$r(1-r)/2$	0	$1-p$	p	n_6
7	$A_2\ B_1 / A_2\ B_1$	$r^2/4$	p^2	$2p(1-p)$	$(1-p)^2$	n_7
8	$A_2\ B_2 / A_2\ B_1$	$r(1-r)/2$	0	p	$1-p$	n_8
9	$A_2\ B_2 / A_2\ B_2$	$(1-r)^2/4$	0	0	1	n_9

p is defined as r_A / r where r_A is the recombination fraction between the left marker 1 and the putative QTL and r is the (known) recombination fraction between the left marker 1 and the right marker 2. C equals $r^2/[r^2+(1-r)^2]$.
Further assumptions are:
—that the possibility of a double recombination event (i.e. crossing over) will be ignored and
—that single crossing over happens independently from one another.
N is the number of plants in the sample F_2 generation and $n_1,....,n_9$ are the number of plants of this generation bearing the combinations of markers 1 to 9.
The conditional probabilities of the QTL genotypes given marker genotypes of Table 15.1 enable us to calculate a matrix $\mathbf{P}=(p_{ji})$ (with dimension $(n \times 3)$, $j=1,2,...,n$ and $i=1,2,3$) for any sample of n plants whose markers are genotyped. This matrix contains the p_{ji} in dependence of the marker genotype of each plant

15. MAXIMUM LIKELIHOOD ESTIMATES

of the sample. The aim is to find Maximum Likelihood estimates of QTL effects including their estimated standard errors.

2. F_2-Generation Model

The genetic model for one QTL represents the relation between a 'genotypic value' G and some genetic parameters β_0, a and d:

$$G = \begin{bmatrix} G_2 \\ G_1 \\ G_0 \end{bmatrix} = \begin{bmatrix} 1 \\ 1 \\ 1 \end{bmatrix} \beta_0 + \begin{bmatrix} 1 & -\frac{1}{2} \\ 0 & \frac{1}{2} \\ -1 & -\frac{1}{2} \end{bmatrix} \begin{bmatrix} a \\ d \end{bmatrix} = \mathbf{1}_{3 \times 1} \beta_0 + \mathbf{DE} \quad (1)$$

Here, β_0 is a joint value of the genetic model and a and d are additive and dominance effects of QTL in the F_2-population. It is possible to calculate unique solutions of the genetic parameters in dependence of the genotypic values and frequencies of genotypes Q_1/Q_1, Q_1/Q_2 and Q_2/Q_2 of the QTL.
$\mathbf{D}=(D_1, D_2)$, where D_1 represents the status of the additive parameter and D_2 represents the status of the dominance effect.
For the following QTL mapping data y_j (j=1,2,...,n) is the investigated trait value of plant j,
$\mathbf{X}_j$ (j=1,2,...,n) is a vector which contains data for the genetic markers and other explanatory variables, and the following assumptions can be made:
—there is no interaction (that is no epistasis) between QTLs
—there is no interference in crossing over
—there is only one QTL in the testing interval.

A statistical composite interval mapping model can be constructed on the basis of the genetic model:

$$y_j = ax_j^* + dz_j^* + X_j\beta + \varepsilon_j \quad (2)$$

Here,
y_j is the trait value of the plant j (j=1,2,...,n),
a and d are additive and dominance effects of the putative QTL,
β is a partial regression coefficient vector of dimension k that contains the mean β_0 of the genetic model,
X_j is a subset of $\mathbf{X}_j$ that contains chosen marker and variable information and $\varepsilon_j \sim N(0, \sigma^2)$.

x_j^* and z_j^* are discrete random effects with

$$x_j^* = \begin{cases} 1 & Q_1/Q_1 \\ 0 \text{ if the QTL is } & Q_1/Q_2 \\ -1 & Q_2/Q_2 \end{cases}$$

$$\text{and } z_j^* = \begin{cases} 1/2 & \text{if the QTL is } Q_1/Q_2 \\ -1/2 & \text{otherwise} \end{cases}, j=1,2,\ldots,n.$$

The realizations of the putative QTL in plant j are unknown. Thus only the probability distribution of the realizations of the discrete random effects can be given in dependence of the conditional probabilities of the QTL genotypes given marker genotypes for plant j (called p_{ji}, with $j=1,2,\ldots,n$, $i=1,2,3$):

$$g_j(x_j^*, z_j^*) = \begin{cases} p_{j1} & \text{if } x_j^* = 1 \text{ and } z_j^* = -1/2 \\ p_{j2} & \text{if } x_j^* = 0 \text{ and } z_j^* = 1/2 \\ p_{j3} & \text{if } x_j^* = -1 \text{ and } z_j^* = -1/2 \end{cases}$$

This is the distribution of the QTL genotype specified by x_j^* and z_j^*.

Now, it is possible to give a Likelihood function for a sample of n individuals and for the parameter vector $\theta = (a, d, \beta, \sigma^2)$:

$$L(\theta|Y, X) = \prod_{j=1}^{n} \left[\sum_{i=1}^{3} p_{ji} f(y_j; \mu_{ji}, \sigma^2) \right]$$

with

$$\mu_{j1} = a - d/2 + X_j \beta$$
$$\mu_{j2} = d/2 + X_j \beta$$
$$\mu_{j3} = -a - d/2 + X_j \beta$$

and f is the normal density of y_j with expectation value μ_{ji} (i=1,2,3 and j=1,2,..,n) and variance σ^2. Another situation, where it is impossible to measure effects directly, is given by Markus et al. (1999).

3. Parameter Estimation by EM Algorithm

The QTL genotypes can be considered as missing values. Now it is possible to define a data set $Y_{mis}=(y_{(mis,j)})$, (with j=1,2,...,n) of "missing data" for the QTL genotypes, and a data set $Y_{obs}=(y_{(obs,j)})$ (with j=1,2,...,n) for the observed values y_j and the marker information (cofactor vectors X_j, j=1,2,...,n). We contemplate a hypothetical complete-data set called $Y_{com}=(Y_{obs}, Y_{mis})$. In such a situation the so-called EM algorithms for Maximum Likelihood estimation of the parameters of the statistical model can be used.

15. MAXIMUM LIKELIHOOD ESTIMATES

Consider the random variable vector Y_{com} of the complete-data set with density function $f(Y_{com} | \theta)$ and $\theta \in \Theta \subseteq \mathbb{R}^d$. If Y_{com} contained only observed values, the objective way to estimate the parameters would be to maximize the complete-data log-likelihood function of θ:

$$l(\theta | Y_{com}) \propto \ln f(Y_{com} | \theta).$$

Unfortunately, Y_{com} contains the not observable missing values Y_{mis}. If we assume that the missing data in Y_{mis} are missing at random, than the log-likelihood for θ is:

$$l_{obs}(\theta | Y_{obs}) \propto \ln \int f(Y_{com} | \theta) dY_{mis}.$$

Now in most practical applications (including the here-described situation) it is very complicated to maximize this log-likelihood-function.
The EM algorithm solves this problem of maximizing l_{obs} by iteratively maximizing $l(\theta | Y_{com})$.
For each iteration, the EM algorithm has two steps, the E-step and the M-step.
Using appropriate starting values for $\theta^{(0)}$,
—**the (t+1) E-step** finds the conditional expectation of the complete data log-likelihood with respect to the conditional distribution of Y_{mis} given Y_{obs} and the parameter $\theta^{(t)}$:

$$Q\!\left(\theta | \theta^{(t)}\right) = \int l\!\left(\theta | Y_{com}\right) f\!\left(Y_{mis} | Y_{obs}, \theta = \theta^{(t)}\right) dY_{mis}$$

This is a function of θ for fixed Y_{obs} and fixed $\theta^{(t)}$.

—**The (t+1) st M-step** calculates a maximum $\theta^{(t+1)}$ for $Q\!\left(\theta | \theta^{(t)}\right)$, so that

$$Q\!\left(\theta^{(t+1)} | \theta^{(t)}\right) \geq Q\!\left(\theta | \theta^{(t)}\right), \; \forall \theta \in \Theta.$$

Under certain restrictions (Dempster, Laird, & Rubin, 1977), the sequence of estimates of the iterations steps of the EM algorithm converges against a (global or local) maximum of l_{obs}. Obviously, depending on the chosen starting values (and the used restrictions) it is possible that in some applications a stationary value of l_{obs} is found, but very often the EM algorithm is able to find a maximum.
For the F_2-generation situation and the models (1) and (2) the observed data $(y_{(obs,j)})$ given the missing data $(y_{(mis,j)})$ are normally distributed:

$$f\!\left(y_{(obs,j)} | \theta, X_j, x_j^*, z_j^*\right) \sim N\!\left(ax_j^* + dz_j^* + X_j\beta, \sigma^2\right).$$

The conditional density of missing data given specified observations is the above defined density of QTL genotypes $g_j(x_j^*, z_j^*)$. The density of the complete data set ($y_{(com,j)}$) can be considered as the likelihood-function and is defined as:

$$L(\theta|y_{(com)}) = \prod_{j=1}^{n} f(y_{(obs,j)}|\theta, X_j, x_j^*, z_j^*) g_j(x_j^*, z_j^*)$$

Now the conditional expectation of the complete data log-likelihood with respect to the conditional distribution of Y_{mis} given Y_{obs} and the parameter $\theta^{(t)}$ is in the **E-step of the EM algorithm**:

$$Q(\theta|\theta^{(t)}) = \int \ln L(\theta|Y_{com}) f(Y_{mis}|Y_{obs}, \theta = \theta^{(t)}) dY_{mis}$$

$$= \int \ln \left[\prod_{j=1}^{n} f(y_j; \mu_{ji}, \sigma^2) g_j(x_j^*, z_j^*) \right] f(Y_{mis}|Y_{obs}, \theta = \theta^{(t)}) dY_{mis}$$

$$= \sum_{j=1}^{n} \sum_{i=1}^{3} \ln[f(y_j; \mu_{ji}, \sigma^2) p_{ji}] \pi_{ji}^{(t)}$$

where $\pi_{ji}^{(t)} = \dfrac{p_{ji} f(y_j; \mu_{ij}^{(t)}, \sigma^{2(t)})}{\sum_{v=1}^{3} p_{jv} f(y_j; \mu_{jv}^{(t)}, \sigma^{2(t)})}$ is the posterior probability of the QTL genotype and the conditional distribution of Y_{mis} given Y_{obs} is given by $f(y_{(mis,j)} | y_{(obs,j)}, \theta = \theta^{(t)}) =$

$$\dfrac{f(y_{(obs,j)}|y_{(mis,j)}) \cdot f(y_{(mis,j)})}{\sum_{l=1}^{3} f(y_{(obs,j)}|y_{(mis,j),l\,fix}) \cdot f(y_{(mis,j),l\,fix})} = \dfrac{g_j(x_j^*, z_j^*) f(y_j; \mu_j, \sigma^2)}{\sum_{v=1}^{3} p_{ji} f(y_j; \mu_{jv}, \sigma^2)}.$$

For the **M-step of the EM algorithm** Q should be maximized.
Setting the partial derivatives of Q equal to zero, we get the matrix equation and explicitly:

$$A\theta^{*(t+1)} = b, \text{ with } \theta^{*(t+1)} = (a^{(t+1)}, d^{(t+1)}, \beta_0^{(t+1)}, \dots, \beta_{k-1}^{(t+1)})^T.$$

15. MAXIMUM LIKELIHOOD ESTIMATES

$$\begin{bmatrix} \sum_{j=1}^{n}\left(\pi_{j1}^{(t)}+\pi_{j3}^{(t)}\right) & \frac{1}{2}\sum_{j=1}^{n}\left(\pi_{j3}^{(t)}-\pi_{j1}^{(t)}\right) & \cdots & \sum_{j=1}^{n}\left(\pi_{j1}^{(t)}-\pi_{j3}^{(t)}\right)X_{jk} \\ \frac{1}{2}\sum_{j=1}^{n}\left(\pi_{j3}^{(t)}-\pi_{j1}^{(t)}\right) & \frac{1}{4}\sum_{j=1}^{n}\left(\pi_{j1}^{(t)}+\pi_{j2}^{(t)}+\pi_{j3}^{(t)}\right) & \cdots & \frac{1}{2}\sum_{j=1}^{n}\left(-\pi_{j1}^{(t)}+\pi_{j2}^{(t)}-\pi_{j3}^{(t)}\right)X_{jk} \\ \vdots & \vdots & & \vdots \\ \sum_{j=1}^{n}\left(\pi_{j1}^{(t)}-\pi_{j3}^{(t)}\right)X_{jk} & \frac{1}{2}\sum_{j=1}^{n}\left(-\pi_{j1}^{(t)}+\pi_{j2}^{(t)}-\pi_{j3}^{(t)}\right)X_{jk} & \cdots & \sum_{j=1}^{n}X_{jk}X_{jk} \end{bmatrix} \begin{pmatrix} a^{(t+1)} \\ d^{(t+1)} \\ \beta_0^{(t+1)} \\ \vdots \\ \beta_{k-1}^{(t+1)} \end{pmatrix} =$$

$$\begin{pmatrix} \sum_{j=2}^{n}\left(\pi_{j1}^{(t)}-\pi_{j3}^{(t)}\right)y_j \\ \frac{1}{2}\sum_{j=1}^{n}\left(-\pi_{j1}^{(t)}+\pi_{j2}^{(t)}-\pi_{j3}^{(t)}\right)y_j \\ \vdots \\ \sum_{j=1}^{n}X_{jk}y_j \end{pmatrix}$$

—It is possible to calculate the parameter estimator of the first (k+1) parameter by sol-ving $A\theta^{*(t+1)} = b$ and to use these estimators for calculating $\sigma^{2(t+1)}$ with

$$\sigma^{2(t+1)} = \frac{1}{n}\sum_{j=1}^{n}\left[\left(y_j - \mu_{j1}^{(t+1)}\right)^2 \pi_{j1}^{(t)} + \left(y_j - \mu_{j2}^{(t+1)}\right)^2 \pi_{j2}^{(t)} + \left(y_j - \mu_{j3}^{(t+1)}\right)^2 \pi_{j3}^{(t)}\right].$$

The parameter estimator vector of the (t+1)-th iteration step of the EM algorithm is

$$\theta^{(t+1)} = \left(a^{(t+1)}, d^{(t+1)}, \beta_0^{(t+1)}, \ldots, \beta_{k-1}^{(t+1)}, \sigma^{(t+1)}\right)^T.$$

Emrich and Urfer (1999) give the asymptotic variance-covariance matrix of these ML-estimates.

The EM algorithm is widely used in different applications. Selinski et al. (2000) used this method for the estimation of toxicokinetic parameters for the risk assessment of potential harmful chemicals. Also Gilberg et al. (1999) applied a modified EM algorithm for parameter estimation in heteroscedastic nonlinear models with random effects. In a recent paper Karlis and Xekalaki (1999) showed how the EM algorithm for mixtures of the one-parameter exponential family can be substantially improved. A modified version of their iterative scheme has already been applied to Minimum Hellinger distance estimation for finite Poisson mixtures by Karlis and Xekalaki (1998).

4. Outlook

In this work we have described a method for the estimation of genetic parameters in an F_2-population. In a forthcoming paper examples will be published showing that the EM algorithm and the ECM algorithm as described by Kao and Zeng may or may not converge against different maxima.

Melchinger et al. (1998) evaluated testcross progenies of 344 F_3 lines in combination with two unrelated testers plus additional testcross progenies from an independent but smaller sample of 107 F_3 lines from the same cross in combination with the same two testers for grain yield and four other important agronomic traits. For a more detailed statistical analysis of this data set A.E. Melchinger and H.F. Utz from the Institute of Plant Breeding, Seed Science and Population Genetics, University of Hohenheim, provided plant height measurements of an F_2 –population of maize, which were genotyped for a total of 89 marker loci. This data set is based on so called adjusted means. Urfer et al. (1999) presented several ways to calculate adjusted means in α-designs.

The aim of our further statistical approach is to find Maximum Likelihood estimates for QTL locations and effects including their estimated standard errors using the described and further methods. Recently, Kao, Zeng and Teasdale (1999) presented a new statistical approach for interval mapping, called multiple marker interval mapping (MIM). It uses multiple marker intervals simultaneously to fit multiple putative QTLs directly in the model for mapping QTLs. Here the ECM algorithm is used as well. To integrate our estimation procedure in this approach seems to be a promising field for further statistical research.

Acknowledgments

We thank the German Research Foundation (DFG) for financial support of the Graduate College and the Collaborative Research Centre at our Department of Statistics.

5. References:

Dempster, A. P., Laird, N. M. and Rubin, D. B. (1977). Maximum Likelihood from Incomplete Data via the EM Algorithm. *J. R. Statist. Soc.*, B 39, 1-38.

Emrich, K. and Urfer, W., (1999). Estimation of genetic parameters using molecular markers and EM algorithms. Technical Report 48/1999. Department of Statistics, University of Dortmund. Available from the world wide web:http://www.statistik.uni-dortmund.de/sfb475/sfblit.htm

Gilberg, F., Urfer, W. and Edler, L., (1999). Heteroscedastic nonlinear regression models with random effects and their application to enzyme kinetic data. *Biometrical Journal*, 41, 543-557.

Jansen, R.C., (1996). A general Monte Carlo method for mapping multiple quantitative trait loci. *Genetics*, 142, 305-311.

Kao, C.-H., and Zeng, Z.-B., (1997). General formulas for obtaining the MLE's and the asymptotic variance-covariance matrix in mapping quantitative trait loci when using the EM algorithm. *Biometrics*, 53, 653-665.

Kao, C.-H., Zeng, Z.-B. and Teasdale, R. D., (1999). Multiple Interval Mapping for Quantitative Trait Loci. *Genetics*, 152, 1203-1216.

Karlis, D. and Xekalaki, E. (1998). Minimum Hellinger distance estimation for Poisson mixtures. *Computational Statistics and Data Analysis*, 29, 81-103.

Karlis, D. and Xekalaki, E. (1999). Improving the EM algorithm for mixtures. *Statistics and Computing*, 9, 303-307.

Márkus, L., Berke, O., Kovacs, J. and Urfer, W. (1999). Spatial prediction of the intensity of latent effects governing hydrogeological phenomena. *Environmetrics*, 10, 633-654.

Melchinger, A.E., Utz, H.F. and Schön, C.C., (1998). Quantitative trait locus (QTL) mapping using different testers and independent populaton samples in maize reveals low power of QTL detection and large bias in estimates of QTL effects. *Genetics*, 149, 383-403.

Selinski, S., Golka, K., Bolt, H.M. and Urfer, W., (2000). Estimation of toxicokinetic parameters in population models for inhalation studies with ethylene. *Environmetrics*, 11, 479- 495.

Stephens, D.A. and Fisch, R.D., (1998). Bayesian analysis of quantitative trait locus data using reversible jump Markov chain Monte Carlo. *Biometrics*, 54, 1334-1347.

Urfer, W. Mejza, S. and Hering, F. (1999). Quantitative trait loci mapping in plant genetics by α - design experiments and molecular genetic marker systems. Technical Report 34/1999, University of Dortmund. Availabe from the world wide web: http://www.statistik.uni-dortmund.dc/sfb475/sfblit.htm.

Received: January 2000

16
A PREDICTIVE MODEL EVALUATION AND SELECTION APPROACH — THE CORRELATED GAMMA RATIO DISTIRIBUTION

Evdokia Xekalaki, John Panaretos & Stelios Psarakis
Department of Statistics
Athens University of Economics & Business, Greece

1. Introduction

Evaluating the forecasting potential of a model before it can be used for planning and decision making has been the concern of many statistical workers. A number of evaluation techniques has thus been considered and much theory has been developed, especially for nested models based mainly on goodness of fit considerations.

Predictive evaluation appears to have received less attention, despite the fact that the predictive ability of a model is a very important characteristic of the model. Xekalaki and Katti (1984) introduced an evaluation scheme of a sequential nature that can be used for models that are not necessarily nested. It is based on the idea of scoring rules for rating the predictive behavior of competing models in which the researcher's subjectivity plays an important role. Its effect is reflected through the rules according to which the performance of the model is scored and rated. (see, also Panaretos et al., 1997, Psarakis, 1993, Psarakis & Panaretos, 1990).

Model comparison problems have also attracted much interest. The selection procedures that have been developed are mainly based on criteria for testing the null hypothesis that one model is valid against an alternative hypothesis that another model is valid. Such testing procedures lead to the selection of one of two competing models. The problem of testing whether two models can be considered as *"equivalent"* in some sense requires a different hypothesis formulation and has only been approached indirectly through the concept of encompassing (see, e.g., Gouriéroux et al., 1993, Gouriéroux & Monfort, 1996) and through asymptotic results based on the change in likelihood.

In this chapter, an evaluation method is proposed that is based on Xekalaki and Katti's idea of using a scoring rule but is free of the element of subjectivity. In particular, a scoring rule is suggested to rate the behavior of a linear forecasting model for each of a series of n points in time. A final rating which embodies the step-by-step scores is then used as a statistic for testing the predictive adequacy of the model. The problem of comparative evaluation is also considered and a test procedure is suggested for testing whether two linear models that are not necessarily nested can be considered to be "*equivalent*" in their predictive abilities. In this case, a distribution which is a generalized form of the F distribution arises as the distribution of the sample statistic is considered. This distribution and the scoring rule associated with it are used for comparing two linear models on real data. In particular, in section 2, the regression model setting considered in the sequel is presented and the scheme suggested for evaluating the predictive ability of a linear model is described. Section 3 deals with the problem of comparatively evaluating two competing linear models in their predictive abilities. The distribution of the test statistic used is derived and studied is sections 4 and 5 while selected percentage points of it are provided in the Appendix. The procedure is illustrated on several crop yield data sets (section 6).

2. Rating the Predictive Ability of a Linear Model

Consider the linear model
$$Y_t = X_t \beta + \varepsilon_t, \quad t = 0, 1, 2, \ldots$$
where Y_t is an $\ell_t \times 1$ vector of observations on the dependent random variable, X_t is an $\ell_t \times m$ matrix of known coefficients $(\ell_0 > m, |X_t'X_t| \neq 0)$, β is an $m \times 1$ vector of regression coefficients and ε_t is an $\ell_t \times 1$ vector of normal error random variables with $E(\varepsilon_t)=0$ and $V(\varepsilon_t)=\sigma^2 I_t$. Here it is the $\ell_t \times \ell_t$ identity matrix. Therefore, a prediction for the value of the dependent random variable for time t+1 will be given by the statistic
$$\hat{Y}_{t+1}^0 = X_{t+1}^{0'} \hat{\beta}_t,$$
where $\hat{\beta}_t = (X_t'X_t)^{-1} X_t' Y_t$ is the least squares estimator of β at time t and X_{t+1}^0 is an $m \times 1$ vector of values of the regressors at time $t+1, t = 0, 1, 2, \ldots$ Obviously,
$$X_{t+1} = \begin{bmatrix} X_t \\ X_{t+1}^{0'} \end{bmatrix} \quad \text{and} \quad Y_{t+1} = \begin{bmatrix} Y_t \\ Y_{t+1}^0 \end{bmatrix}$$
are of dimension $\ell_{t+1} \times m$ and $\ell_{t+1} \times 1$ respectively, where $\ell_{t+1} = \ell_t + 1, \; t = 0, 1, 2, \ldots$.

The predictive behavior of the model would naturally be evaluated by a measure that would be based on a statistic reflecting the degree of agreement of the observed actual value $\hat{Y}_{t+1}^0$ to the predicted value $\hat{Y}_{t+1}^0$. Such a statistic may be the statistic $|r_{t+1}|$, where

$$r_{t+1} = \frac{\hat{Y}_{t+1}^0 - Y_{t+1}^0}{S_t \sqrt{1 + X_{t+1}^{0'}(X_t'X_t)^{-1} X_{t+1}^0}}, \quad t = 0, 1, \ldots \quad (1)$$

Obviously, $|r_{t+1}|$ is merely an estimate of the standardized distance between the predicted and the observed value of the dependent random variable when σ^2 is estimated on the basis of the preceding ℓ_t observations available at time t. S_t^2 is given by

i.e., $$S_t^2 = \frac{(Y_t - X_t\hat{\beta}_t)'(Y_t - X_t\hat{\beta}_t)}{(\ell_t - m)}, \quad t = 0, 1, 2, \ldots,$$

So, a score based on $|r_{t+1}|$ can provide a measure of the predictive adequacy of the model for each of a series of n points in time. Then, as a final rating of the model one can consider the average of these scores, or any other summary statistic that can be regarded as reflecting the forecasting potential of the model.

In the sequel, we consider using r_t^2 as a scoring rule to rate the performance of the model at time t for a series of n points in time, (t =1, 2, ..., n) and we define

$$R_n = \sum_{t=1}^{n} r_t^2 \Big/ n \qquad (2)$$

the average of the squared recursive residuals, to be the final rating of the model.

It has been shown (Brown, et al., 1975, Kendall et al., 1983) that if ε_t is a vector of normal error variables with $E(\varepsilon_t)=0$ and $V(\varepsilon_t)=\sigma^2 I_t$, the quantities

$$W_{t+1} = \frac{\hat{Y}_{t+1}^0 - Y_{t+1}^0}{\sqrt{1 + X_{t+1}^{0'}(X_t'X_t)^{-1}X_{t+1}^0}}, \quad t = 0, 1, 2, \ldots$$

are independently and identically distributed normal variables with mean 0 and variance σ^2. Then, according to Kotlarski's (1966) characterization of the normal distribution by the t distribution, the quantities $r_{t+1} = w_{t+1}/s_t$, $t = 0, 1, 2, \ldots$ constitute a sequence of independent t variables with $\ell_t - m$ degrees of freedom, $t = 0, 1, 2, \ldots$. Hence, by the assumptions of the model considered and for large ℓ_0, the variables r_{t+1}, $t = 0, 1, 2, \ldots$ constitute a sequence of approximately standard normal variables which are mutually independent. This implies that

$$nR_n = \sum_{t=1}^{n} r_t^2$$

is a chi-square variable with n degrees of freedom.

3. Comparative Evaluation of the Predictive Ability of Two Linear Models With the Use of a Generalized Form of the F Distribution

Consider now A and B to be two competing linear models that have been used for prediction purposes for a number n_1 and n_2 of years, respectively. A null hypothesis that is interesting to test is whether two models have

16. A PREDICTIVE MODEL EVALUATION

"*equivalent*" forecasting abilities. This is a hypothesis that can be defined only implicitly, but it exists as a mathematical entity. The closest description of it is "*H_0: models A and B have equal mean squared prediction errors.*" This is a hypothesis that can be tested formally using conventional methods, in all cases in which neither, one, or both models are correctly specified using the average standardized distances between the observed value of the dependent variable and its predicted values by models A and B. Then, a decision on whether models A and B are "*equivalent*" in their predictive ability would naturally be based on the ratio of the average scores of the two models as given by the statistic

$$R_{n_1,n_2} = \frac{R_{n_1}(A)}{R_{n_2}(B)} \qquad (3)$$

where $R_{n_1}(A)$, $R_{n_2}(B)$, are given by (2) for $n=n_1$ and $n=n_2$ and refer to model A and model B, respectively.

For large ℓ_{t_1}, ℓ_{t_2} the distribution of the statistic R_{n_1,n_2} can be approximated by the F distribution with n_1 and n_2 degrees of freedom whenever the ratings of the two models are independent. Hence, values of R_{n_1,n_2} in the right tail of the F distribution with n_1 and n_2 degrees of freedom will indicate a higher performance by model A.

However, under the conditions of the problem, the assumption of independence does not seem to be satisfied.

Determining the exact distribution of R_{n_1,n_2} in the case of dependent ratings would, however, be desirable as in practice data on ratings are often matched. (In the latter case, $n_1=n_2=n$.)

Kotlarski (1964) has shown that, under certain conditions, the quotient X/Y, where X,Y are positive valued random variables not necessarily independent, follows the F distribution. According to Kotlarski (1964), a necessary and sufficient condition for the ratio of two variables to follow an F distribution can be established through the form of the Mellin transform of their joint distribution. In particular, Kotlarski (1964) has shown that if Ψ is the set of joint distribution functions F(x,y) of two not necessarily independent positive valued random variables X and Y, whose quotient X/Y follows the F distribution with parameters p_1 and p_2, then the following result holds.

Theorem (Kotlarski, 1964): For a distribution function F(x,y) to belong to the set Ψ it is necessary and sufficient that its Mellin transform

$$h(u,v) = \int_0^\infty \int_0^\infty x^u y^v dF(x,y)$$

satisfies the condition

$$h(u,-u) = \frac{\Gamma(p_1+u)}{\Gamma(p_1)} \frac{\Gamma(p_2-u)}{\Gamma(p_2)}.$$

For our problem, consider the random variables $X_i = r_i(A)$, $Y_i = r_i(B)$, $i=1, 2,\ldots, n$ obtained from (1) for model A and model B respectively. Each of the variables X_i, Y_i follows the standard normal distribution. The joint distribution is therefore the bivariate standard normal distribution with a correlation coefficient denoted by ρ. Under these conditions, the joint distribution of the random variables

$$X = \frac{\sum_{i=1}^{n} X_i^2}{n} = R_n(A) \quad \text{and} \quad Y = \frac{\sum_{i=1}^{n} Y_i^2}{n} = R_n(B)$$

is Kibble's (1941) bivariate Gamma distribution as defined by the probability density function

$$f(x,y) = \frac{\rho^{-(k-1)}}{\Gamma(k)(1-\rho^2)}(xy)^{\frac{k-1}{2}} e^{-\frac{x+y}{1-\rho^2}} I_{k-1}\left[\frac{2\rho\sqrt{xy}}{1-\rho^2}\right], \quad (4)$$

where $k = n/2$ and $I_k(x)$ is the modified Bessel function of the first kind of order k given by (see Abramowitz & Stegun, 1972)

$$I_k(x) = \sum_{i=0}^{\infty} \left(\frac{x}{2}\right)^{k+2i} \frac{1}{\Gamma(i+1)\Gamma(i+k+1)}. \quad (5)$$

Therefore,

$$f(x,y) = \frac{\rho^{-(k-1)}}{\Gamma(k)(1-\rho^2)} e^{-\frac{x+y}{1-\rho^2}} \sum_{i=0}^{\infty} \left(\frac{\rho}{1-\rho^2}\right)^{k+2i-1} \frac{x^{\frac{k-1}{2}+\frac{k-1}{2}+i} y^{\frac{k-1}{2}+\frac{k-1}{2}+i}}{\Gamma(i+1)\Gamma(i+k)},$$

So, finally, the probability density function of the bivariate gamma distribution of $(R_n(A), R_n(B))$ is given by

$$f(x,y) = \frac{e^{-\frac{x+y}{1-\rho^2}}}{\Gamma(k)(1-\rho^2)^k} \sum_{i=0}^{\infty} \frac{(\rho/(1-\rho^2))^{2i}}{\Gamma(i+1)\Gamma(i+k)} (xy)^{k-1+i}$$

To determine whether an F form can be deduced for the distribution of $R_{n,n}$, one needs to examine if Kotlarski's theorem applies for the joint distribution of $R_n(A)$, $R_n(B)$.

For Kibble's bivariate Gamma distribution, we obtain, by the definition of the Mellin transform
$h(u,v) = E(X^u Y^v)$

$$= \frac{(1-\rho^2)^{-k}}{\Gamma(k)} \sum_{i=0}^{\infty} \left(\frac{\rho}{1-\rho^2}\right)^{2i} \frac{1}{\Gamma(i+1)\Gamma(i+k)} \int_0^{\infty}\int_0^{\infty} e^{-\frac{x+y}{1-\rho^2}} x^{u+k-1+i} y^{v+k+i-1} dx dy.$$

Definition by I, the double integral in the right-hand side of the above relationship, we have

$$I = \frac{\Gamma(u+k+i)\,\Gamma(v+k+i)}{(1-\rho^2)^{-(u+v+2k+2i)}}.$$

This, in turn, implies that

$$h(u,v) = \frac{(1-\rho^2)^{-k+u+v+2k}}{\Gamma(k)} \sum_{i=0}^{\infty} \frac{\rho^{2i} \Gamma(u+k+i) \Gamma(v+k+i)}{i! \Gamma(k+i)} =$$

$$= \frac{(1-\rho^2)^{u+v+k}}{\Gamma(k)} \frac{\Gamma(k+u)\Gamma(k+v)}{\Gamma(k)} \sum_{i=0}^{\infty} \frac{(k+u)_{(i)}(k+v)_{(i)}}{k_{(i)}} \frac{\rho^{2i}}{i!},$$

or, equivalently that

$$h(u,v) = \frac{\Gamma(k+u)}{\Gamma(k)} \frac{\Gamma(k+v)}{\Gamma(k)} (1-\rho^2)^{u+v+k} {}_2F_1(k+u, k+v; k; \rho^2), \quad (6)$$

where

$${}_2F_1(a,b;c;z) = \sum_{r=0}^{\infty} \frac{a_{(r)} b_{(r)}}{c_{(r)}} \frac{z^r}{r!}$$

is the hypergeometric series with $\alpha_{(r)}$ denoting the ascending factorial (see Abramowitz & Stegun, 1972).

One can see that the Mellin transform of Kibble's distribution given (6) does not satisfy the conditions of Theorem 1. Hence, the quotient $R_n(A)/R_n(B)$ does not follow the F distribution when $R_n(A)$ and $R_n(B)$ are dependent.

In the next section, it is shown that the distribution of $R_{n,n}$ is a generalized form of the F distribution.

4. The Distribution of the Ratio X/Y When X and Y Follow Kibble's Bivariate Gamma Distribution

It is known that if X and Y are dependent random variables, the distribution function of Z=X/Y is given by

$$F_Z(z) = P(X/Y \le z) = \int_0^{\infty} P(X \le zy | Y = y) f_Y(y) dy,$$

where $F_U(\cdot)$ and $f_U(\cdot)$ denote the distribution function and the probability density function of a random variable U respectively.

Then, the density function of the quotient Z=X/Y can be written as

$$f_Z(z) = \int_0^{\infty} f_{X|Y=y}(zy) f_Y(y) dy = \int_0^{\infty} \frac{f_{X,Y}(zy,y)}{f_Y(y)} y \, f_Y(y) dy$$

$$= \int_0^{\infty} y \, f_{X,Y}(zy,y) dy .$$

This leads to

$$f_{X/Y}(z) = \int_0^\infty y\, f_{X,Y}(zy,y)\,dy$$

$$= \frac{1}{(1-\rho^2)^k \Gamma(k)} \sum_{i=0}^\infty \frac{\rho^{2i}}{(1-\rho^2)^{2i} i!\,\Gamma(i+k)} \int_0^\infty \exp\left(-\frac{zy+y}{1-\rho^2}\right) z^{k+i-1}\, y^{2(k+i)-1}\,dy$$

$$= \frac{z^{k-1}}{(1-\rho^2)^k \Gamma(k)} \sum_{i=0}^\infty \frac{\rho^{2i} z^i}{(1-\rho^2)^{2i} i!\,\Gamma(i+k)} \int_0^\infty \exp\left(-y\frac{z+1}{1-\rho^2}\right) y^{2(k+i)-1}\,dy$$

$$= \frac{z^{k-1}}{(1-\rho^2)^k \Gamma(k)} (1+z)^{-2k} \sum_{i=0}^\infty \frac{\Gamma(2k+2i)}{\Gamma(i+k)} \left(\frac{\rho^2}{(1+z)^2}\right)^i \frac{z^i}{i!}. \qquad (7)$$

Furthermore,

$$\frac{\Gamma(2k+2i)}{\Gamma(k)\Gamma(i+k)} = \frac{\Gamma(2k+2i)\Gamma(2k)\Gamma(k)}{\Gamma(k)\Gamma(i+k)\Gamma(2k)\Gamma(k)} = \frac{(2k)_{(2i)}}{k_{(i)}} [B(k,k)]^{-1}.$$

$$= [B(k,k)]^{-1} \frac{2^{2i}\left[\frac{2k}{2}\right]_{(i)}\left[\frac{2k+1}{2}\right]_{(i)}}{k_{(i)}}$$

Here, we made use of the identities

$$B(\alpha,\beta) = \frac{\Gamma(\alpha)\Gamma(\beta)}{\Gamma(\alpha+\beta)}$$

and

$$\alpha_{(mn)} = n^{nm}\left(\frac{\alpha}{n}\right)_{(m)}\left(\frac{\alpha+1}{n}\right)_{(m)} \cdots \left(\frac{\alpha+n-1}{n}\right)_{(m)}.$$

Letting $\alpha=2k$, $m=i$, $n=2$ one obtains

$$\frac{\Gamma(2k+2i)}{\Gamma(k)\Gamma(i+k)} = \frac{2^{2i}\left[\frac{2k+1}{2}\right]_{(i)}}{B(k,k)}.$$

Hence (7) can be written as

$$f_{X/Y}(z) = (1-\rho^2)^k \frac{z^{k-1}(1+z)^{-2k}}{B(k,k)} \sum_{i=0}^\infty \left[\frac{2k+1}{2}\right]_{(i)} \frac{[4\rho^2(z+1)^{-2}]^i z^i}{i!}$$

$$= (1-\rho^2)^k \frac{z^{k-1}(1+z)^{-2k}}{B(k,k)} \left[1 - 4\frac{\rho^2 z}{(z+1)^2}\right]^{-\frac{2k+1}{2}}$$

Therefore,

$$f_{X/Y}(z) = \frac{(1-\rho^2)^k}{B(k,k)} z^{k-1}(1+z)^{-2k}\left[1 - \left[\frac{2\rho}{z+1}\right]^2 z\right]^{-\frac{2k+1}{2}}. \qquad (8)$$

16. A PREDICTIVE MODEL EVALUATION

The density function in (8) defines the distribution of the quotient X/Y when the joint distribution of (X,Y) is Kibble's bivariate gamma. In the sequel, we refer to this distribution as *the correlated gamma - ratio (CGR) distribution with parameters ρ and k*. (A reparameterized form of this distribution was arrived at by Izawa (1965)).

Note: One can see that in the case where X and Y are independent, whence $\rho=0$, the probability density function of the quotient X/Y takes the form

$$f_{X/Y}(z) = \frac{1}{B(k,k)} z^{k-1}(1+z)^{-2k}.$$

This is the probability density function of the Beta type II distribution with parameters k and R or, equivalently of the F distribution with 2k and 2k degrees of freedom.

5. The t Distribution as a Limiting Case of the Correlated Gamma Ratio Distribution

In the sequel, it is shown that the t distribution can be obtained as a limiting case of the CGR distribution.

Let Z follow the CGR distribution with density function given by (8). Consider the variable

$$T = \frac{\rho}{\sqrt{1-\rho^2}} \frac{Z-1}{Z+1}.$$

Then,

$$F_T(t) = P(T \le t) = P\left(Z \le \frac{\rho + t\sqrt{1-\rho^2}}{\rho - t\sqrt{1-\rho^2}}\right) = F_Z\left(\frac{\rho + t\sqrt{1-\rho^2}}{\rho - t\sqrt{1-\rho^2}}\right),$$

where $-\frac{\rho}{\sqrt{1-\rho^2}} < t < \frac{\rho}{\sqrt{1-\rho^2}}$.

We have therefore, for the probability density function of T that

$$f_T(t) = f_Z\left(\frac{\rho + t\sqrt{1-\rho^2}}{\rho - t\sqrt{1-\rho^2}}\right) \frac{2\rho\sqrt{1-\rho^2}}{\left(\rho - t\sqrt{1-\rho^2}\right)^2},$$

where $-\frac{\rho}{\sqrt{1-\rho^2}} < t < \frac{\rho}{\sqrt{1-\rho^2}}$.

Using (8), this reduces to

$$f_T(t) = \frac{1}{\rho} \frac{2^{1-2k}}{B(k,k)} \left[1 - \left(\frac{\sqrt{1-\rho^2}}{\rho} t\right)^2\right]^{k-1} (1+t^2)^{-\frac{2k+1}{2}},$$

where $-\frac{\rho}{\sqrt{1-\rho^2}} < t < \frac{\rho}{\sqrt{1-\rho^2}}$.

Taking the limit as $\rho \to 1$ we obtain

$$\lim_{\rho \to 1} f_T(t) = \frac{2^{1-2k}}{B(k,k)}\left(1+t^2\right)^{-\frac{2k+1}{2}}, \qquad -\infty < t < +\infty.$$

But this is the probability density function of the t distribution.

In the Appendix, some graphs of the probability density function of the correlated gamma-ratio distribution are provided for different values of k and ρ. Also, Tables A1, A2 and A3 provide percentage points of the distribution for selected values of the parameter k (k=1(1) 30, 40, 50, 60) and of the correlation coefficient ρ (ρ=0.0(0.1) 0.9).

6. An Application to Crop-Yield Data

For the purpose of illustrating the model selection procedure, a problem presented in Xekalaki and Katti (1984), concerning the selection of a linear model among several competing ones considered by the United States Department of Agriculture (USDA) to predict the corn yield for 10 Crop Reporting Districts (CRD 10, 20, ...,100), was re-examined based on several sets of real data for the State of Iowa for the years 1956 to 1980. The competing models use information about the weather conditions (e.g., temperature, rainfall etc.) for the previous time periods as well as general trend factors for predicting the crop yield. A detailed description of the models can be found in Linardis (1998).

The aim of the application is to compare the predictability of these models for every district, using the Correlated Gamma - Ratio distribution.

Let m_A and m_B denote these two models respectively. To compare the two crop yield models we need to test a hypothesis of the form:

H_0: *Models m_A and m_B are of "equivalent" predictive ability* (symbolically, $m_A \sim m_B$)versus an alternative

H_1: *The two models differ in their predictive ability, i.e., m_A is of higher predictive ability (symbolically, $m_A \succ m_B$) or of lower predictive ability (symbolically, $m_A \prec m_B$)*,

where the term "equivalent" is used in the sense defined in section 3.

Rejection of the null hypothesis indicates that one of the models performs differently. With a one-sided alternative, one may proceed in a manner similar to that used when testing for equality of variances via the F-test. The results of testing the predictive equivalence of models m_A and m_B on the crop yield data and considered together with the estimated values of the correlations between the standardized prediction errors for the two models are summarized in Table16.1.

Table 16.1: Results of testing the null hypothesis of predictive equivalence of models m_A and m_B H_0: $m_A \sim m_B$ on the crop yield data of the 10 reporting districts the state of Iowa (n=24).

Crop reporting district	H_1	Sums of squared recursive residuals		$R_{n,n}$	Estimated value of ρ	p-value	model to be selected ("best" model)
		Model m_A $(n R_n(A))$	Model m_B $(n R_n(B))$				
CRD 10	$m_A \succ m_B$	58.844	92.798	0.634	0.803	0.0355	model A
CRD 20	$m_A \succ m_B$	58.681	59.595	0.985	0.908	0.4656	"equivalent"
CRD 30	$m_A \succ m_B$	24.638	35.354	0.697	0.885	0.0337	model A
CRD 40	$m_A \prec m_B$	69.677	66.691	1.044	0.449	0.453	"equivalent"
CRD 50	$m_A \succ m_B$	49.005	51.028	0.961	0.620	0.45	"equivalent"
CRD 60	$m_A \prec m_B$	55.949	32.789	1.706	0.155	0.0963	model B
CRD 70	$m_A \succ m_B$	39.933	49.012	0.815	0.561	0.275	"equivalent"
CRD 80	$m_A \prec m_B$	57.396	52.232	1.098	0.796	0.353	"equivalent"
CRD 90	$m_A \prec m_B$	61.461	41.810	1.470	0.669	0.1068	"equivalent"
CRD 100	$m_A \succ m_B$	46.515	73.943	0.629	0.593	0.0868	model A

From this table, one may see that for six districts, the models are of equivalent predictive ability. Model m_A performs *"better"* in 3 cases while only in one case model m_B is *"superior."*

In all the cases considered, the parameter ρ was estimated from the data as the sample correlation between the standardized prediction errors of the two competing models. The extent to which the use of an estimate of ρ may affect the selection procedure has to be investigated. Of course, asymptotically, it is not expected to have any impact because ρ is estimated consistently. The first investigation results for small to moderate sample sizes are not indicative of any appreciable effect either.

APPENDIX

Table A1: Percentage points of the Correlated Gamma Ratio distribution for α=0.1

$$\int_0^z \frac{\left(1-\rho^2\right)^k}{B(k,k)} t^{k-1}(1+t)^{-2k}\left[1-\left[\frac{2\rho}{t+1}\right]^2 t\right]^{-\frac{2k+1}{2}} dt = 1-\alpha = 0.90$$

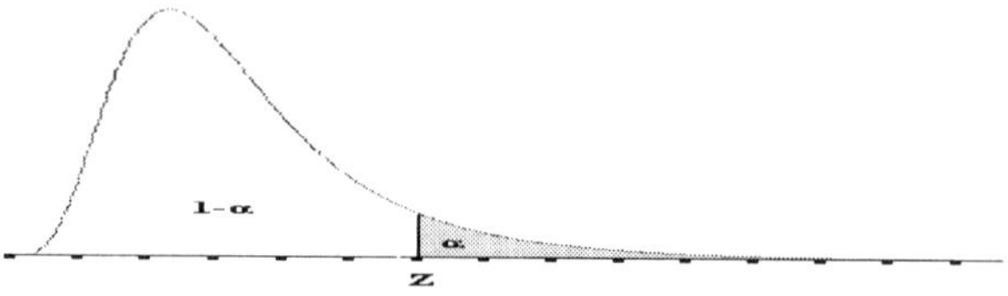

k \ ρ	0.0	0.1	0.2	0.3	0.4	0.5	0.6	0.7	0.8	0.9
1	9	8.93	8.72	8.36	7.85	7.2	6.4	5.45	4.33	3.02
2	4.11	4.08	4.01	3.88	3.71	3.48	3.2	2.85	2.44	1.93
3	3.055	3.04	3.00	2.92	2.81	2.67	2.49	2.27	2.00	1.66
4	2.59	2.58	2.55	2.49	2.41	2.3	2.17	2.00	1.8	1.53
5	2.32	2.31	2.29	2.24	2.18	2.09	1.98	1.84	1.67	1.46
6	2.15	2.14	2.12	2.08	2.02	1.95	1.85	1.74	1.59	1.41
7	2.02	2.01	2.00	1.96	1.91	1.85	1.76	1.66	1.54	1.37
8	1.93	1.92	1.90	1.87	1.83	1.77	1.70	1.61	1.49	1.34
9	1.85	1.846	1.83	1.80	1.76	1.71	1.64	1.56	1.455	1.315
10	1.79	1.785	1.775	1.75	1.71	1.665	1.6	1.525	1.425	1.295
11	1.745	1.74	1.725	1.705	1.67	1.62	1.565	1.49	1.4	1.277
12	1.705	1.70	1.685	1.665	1.63	1.59	1.535	1.465	1.38	1.265
13	1.665	1.664	1.65	1.63	1.60	1.56	1.51	1.44	1.36	1.253
14	1.635	1.63	1.62	1.6	1.57	1.53	1.485	1.423	1.345	1.24
15	1.605	1.604	1.59	1.575	1.546	1.51	1.465	1.405	1.33	1.31
16	1.585	1.58	1.57	1.55	1.525	1.49	1.445	1.39	1.32	1.225
17	1.56	1.553	1.546	1.53	1.505	1.471	1.43	1.376	1.307	1.216
18	1.54	1.535	1.525	1.510	1.486	1.455	1.415	1.364	1.297	1.207
19	1.52	1.519	1.51	1.495	1.471	1.44	1.402	1.351	1.287	1.203
20	1.505	1.504	1.495	1.48	1.456	1.426	1.39	1.341	1.28	1.197
21	1.49	1.489	1.48	1.465	1.44	1.415	1.377	1.331	1.274	1.193
22	1.475	1.474	1.466	1.451	1.43	1.404	1.379	1.323	1.353	1.187
23	1.465	1.460	1.455	1.440	1.567	1.391	1.358	1.315	1.259	1.183
24	1.454	1.450	1.442	1.428	1.408	1.382	1.35	1.306	1.252	1.178
25	1.442	1.44	1.432	1.418	1.4	1.374	1.34	1.3	1.246	1.174
26	1.432	1.43	1.422	1.408	1.39	1.366	1.344	1.292	1.240	1.17
27	1.422	1.42	1.412	1.4	1.382	1.356	1.326	1.286	1.238	1.166
28	1.412	1.410	1.402	1.39	1.372	1.35	1.32	1.28	1.23	1.163
29	1.404	1.402	1.394	1.382	1.366	1.342	1.312	1.274	1.226	1.16
30	1.396	1.394	1.386	1.375	1.358	1.336	1.306	1.27	1.222	1.157
40	1.333	1.332	1.326	1.316	1.302	1.284	1.259	1.228	1.189	1.134
50	1.293	1.291	1.287	1.279	1.267	1.249	1.229	1.203	1.168	1.119
60	1.265	1.264	1.259	1.252	1.24	1.226	1.207	1.183	1.152	

16. A PREDICTIVE MODEL EVALUATION

Table A2: Percentage points of the Correlated Gamma Ratio distribution for α=0.05

$$\int_0^z \frac{(1-\rho^2)^k}{B(k,k)} t^{k-1}(1+t)^{-2k}\left[1-\left[\frac{2\rho}{t+1}\right]^2 t\right]^{-\frac{2k+1}{2}} dt = 1-\alpha = 0.95$$

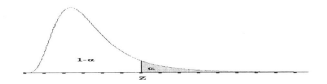

ρ k	0.0	0.1	0.2	0.3	0.4	0.5	0.6	0.7	0.8	0.9
1	19	18.80	18.3	17.4	16.27	14.73	12.84	10.60	8.02	5.04
2	6.39	6.34	6.20	5.97	5.64	5.22	4.7	4.07	3.34	2.46
3	4.284	4.26	4.18	4.04	3.85	3.61	3.31	2.945	2.51	1.97
4	3.44	3.42	3.36	3.27	3.145	2.96	2.74	2.48	2.16	1.76
5	2.98	2.96	2.92	2.84	2.74	2.6	2.43	2.22	1.965	1.64
6	2.687	2.675	2.65	2.57	2.485	2.37	2.23	2.06	1.835	1.56
7	2.49	2.47	2.44	2.39	2.31	2.21	2.09	1.935	1.75	1.51
8	2.335	2.325	2.29	2.25	2.18	2.1	1.985	1.85	1.675	1.46
9	2.22	2.21	2.19	2.14	2.18	2	1.95	1.775	1.63	1.427
10	2.125	2.115	2.095	2.055	2	1.93	1.837	1.725	1.585	1.4
11	2.05	2.04	2.02	1.983	1.935	1.87	1.783	1.677	1.55	1.375
12	1.983	1.977	1.955	1.925	1.876	1.815	1.735	1.635	1.515	1.355
13	1.93	1.922	1.905	1.875	1.83	1.775	1.697	1.605	1.49	1.338
14	1.884	1.876	1.86	1.83	1.787	1.733	1.663	1.577	1.47	1.324
15	1.843	1.835	1.82	1.794	1.752	1.7	1.63	1.552	1.453	1.31
16	1.805	1.798	1.783	1.757	1.72	1.675	1.61	1.527	1.427	1.297
17	1.775	1.767	1.753	1.727	1.697	1.644	1.582	1.508	1.414	1.287
18	1.745	1.74	1.723	1.697	1.667	1.620	1.563	1.493	1.397	1.277
19	1.717	1.711	1.697	1.678	1.644	1.59	1.543	1.472	1.387	1.27
20	1.695	1.69	1.676	1.653	1.624	1.576	1.527	1.46	1.375	1.262
21	1.672	1.667	1.654	1.633	1.604	1.564	1.511	1.447	1.362	1.254
22	1.654	1.647	1.635	1.613	1.584	1.549	1.498	1.434	1.353	1.247
23	1.633	1.629	1.617	1.597	1.567	1.531	1.484	1.424	1.344	1.242
24	1.615	1.612	1.6	1.581	1.553	1.516	1.469	1.412	1.336	1.236
25	1.6	1.596	1.585	1.566	1.54	1.504	1.458	1.401	1.328	1.229
26	1.585	1.581	1.57	1.552	1.526	1.491	1.447	1.390	1.320	1.224
27	1.57	1.566	1.558	1.54	1.514	1.48	1.437	1.383	1.314	1.22
28	1.558	1.556	1.544	1.528	1.502	1.47	1.426	1.374	1.307	1.215
29	1.546	1.543	1.532	1.516	1.492	1.459	1.418	1.367	1.302	1.211
30	1.534	1.531	1.522	1.505	1.482	1.45	1.41	1.359	1.296	1.207
40	1.447	1.445	1.437	1.423	1.404	1.378	1.346	1.303	1.249	1.175
50	1.391	1.390	1.382	1.37	1.355	1.332	1.304	1.267	1.22	1.156
60	1.353	1.35	1.345	1.334	1.319	1.299	1.274	1.241	1.199	

Table A3: Percentage points of the Correlated Gamma Ratio distribution for α=0.01

$$\int_0^z \frac{\left(1-\rho^2\right)^k}{B(k,k)} t^{k-1}(1+t)^{-2k}\left[1-\left[\frac{2\rho}{t+1}\right]^2 t\right]^{-\frac{2k+1}{2}} dt = 1-\alpha = 0.99$$

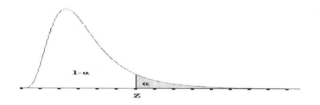

k \ ρ	0.0	0.1	0.2	0.3	0.4	0.5	0.6	0.7	0.8	0.9
1	99	98.10	95.2	90.3	83.5	74.8	64.1	51.7	36.7	20.4
2	15.98	15.84	15.42	14.71	13.72	12.45	10.90	9.05	6.91	4.45
3	8.47	8.40	8.20	7.87	7.40	6.8	6.05	5.17	4.13	2.91
4	6.03	5.99	5.86	5.64	5.34	4.95	4.47	3.89	3.2	2.38
5	4.85	4.82	4.73	4.57	4.34	4.05	3.69	3.25	2.73	2.11
6	4.155	4.13	4.06	3.93	3.75	3.52	3.23	2.88	2.46	1.94
7	3.7	3.68	3.62	3.51	3.36	3.16	2.92	2.62	2.27	1.83
8	3.37	3.36	3.30	3.21	3.08	2.91	2.7	2.45	2.14	1.75
9	3.13	3.12	3.07	2.99	2.87	2.72	2.53	2.31	2.03	1.68
10	2.94	2.93	2.88	2.81	2.705	2.565	2.405	2.2	1.95	1.63
11	2.785	2.775	2.735	2.67	2.575	2.45	2.3	2.11	1.88	1.59
12	2.66	2.65	2.61	2.55	2.465	2.35	2.21	2.04	1.825	1.555
13	2.555	2.545	2.51	2.455	2.375	2.27	2.135	1.975	1.78	1.525
14	2.465	2.455	2.425	2.37	2.295	2.195	2.075	1.925	1.74	1.497
15	2.39	2.38	2.35	2.3	2.23	2.135	2.025	1.88	1.705	1.475
16	2.32	2.31	2.285	2.235	2.17	2.08	1.975	1.84	1.675	1.46
17	2.26	2.25	2.225	2.18	2.117	2.035	1.935	1.805	1.645	1.437
18	2.208	2.195	2.172	2.13	2.07	1.99	1.895	1.773	1.62	1.418
19	2.16	2.15	2.127	2.086	2.03	1.955	1.86	1.744	1.599	1.41
20	2.115	2.105	2.085	2.046	1.994	1.92	1.83	1.72	1.58	1.395
21	2.075	2.07	2.049	2.01	1.956	1.89	1.801	1.695	1.56	1.384
22	2.04	2.034	2.01	1.976	1.925	1.86	1.775	1.675	1.544	1.374
23	2.005	2	1.98	1.946	1.897	1.835	1.754	1.654	1.53	1.364
24	1.978	1.972	1.952	1.918	1.872	1.810	1.732	1.634	1.512	1.352
25	1.95	1.944	1.924	1.892	1.848	1.788	1.712	1.618	1.5	1.344
26	1.924	1.918	1.90	1.868	1.824	1.766	1.694	1.602	1.488	1.336
27	1.9	1.894	1.876	1.846	1.804	1.748	1.676	1.588	1.476	1.328
28	1.878	1.872	1.854	1.826	1.784	1.73	1.66	1.574	1.464	1.32
29	1.856	1.852	1.834	1.806	1.766	1.712	1.645	1.561	1.455	1.314
30	1.838	1.832	1.816	1.788	1.748	1.696	1.632	1.55	1.446	1.308
40	1.69	1.685	1.672	1.65	1.619	1.578	1.525	1.458	1.374	1.259
50	1.597	1.594	1.583	1.565	1.538	1.502	1.456	1.4	1.327	1.229
60	1.536	1.532	1.522	1.506	1.48	1.449	1.409	1.359	1.294	-

The probability density function of the Correlated Gamma Ratio Distribution

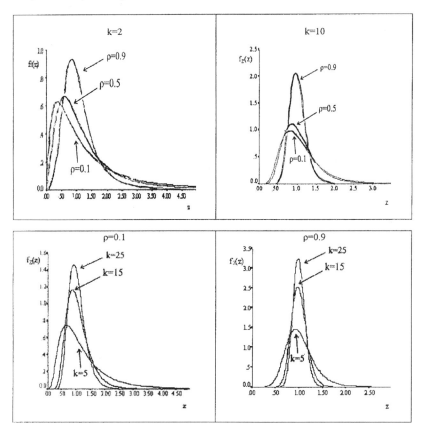

The probability density function of the Correlated Gamma-Ratio distribution for selected values of k and ρ

Acknowledgment

The authors thank Professors D.R. Cox and C.R. Rao for their comments on an earlier version of the chapter.

7. References

Abramowitz, M. and Stegun, I.A. (1972), *Handbook of Mathematical Functions.* New York, Dover.
Brown, R. L., Durbin, J. and Evans, J. M. (1975). Techniques for Testing the Constancy of Regression Relationships Over Time. *Journal of the Royal Statistical Society, B* 37, 149-192.
Gouriéroux, C. and Monfort A. (1996). *Simulation-Based Econometric Methods,* CORE Lectures, Oxford University Press.
Gouriéroux, C., Monfort A., and Renault E. (1993). "Indirect Inference". *Journal of Applied Econometrics,* 8, 85-118.

Izawa, T. (1965). Two or Multi-dimensional Gamma-type Distribution and Its Application to Rainfall Data. *Papers in Meteorology and Geophysics*, XV (3 ~ 4), 167-200.

Kendall, M. Stuart, A. and Ord, J. K. (1983). *The Advanced Theory of Statistics: Vol. 1 (4th edition)*. Griffin, London.

Kibble, W. F. (1941). A Two Variate Gamma Type Distribution. *Sankhya*, 5, 137-150.

Kotlarski, I. (1964). On Bivariate Random Variables Where the Quotient of their Coordinates Follows Some Known Distributions. *Annals of Mathematical Statistics*, 1673-1684.

Kotlarski, I. (1966). On Characterizing the Normal Distribution by the Student's Law. *Biometrika*, 53, 603-606.

Linardis A. (1998). A Comparative Study of Test Procedures Used in Assessing the Forecasting Ability of Linear Models on Crop Yield Data. *M.Sc. Thesis, Department of Statistics, Athens University of Economics and Business.*

Panaretos, J., Psarakis, S. and Xekalaki, E. (1997). The Correlated Gamma-Ratio Distribution in Model Evaluation and Selection. Technical Report no. 33, Department of Statistics, Athens University of Economics and Business.

Psarakis, S. (1993). Methods of Evaluating the Predictive Ability of Linear Models. *Ph.D. Thesis, Department of Statistics, Athens University of Economics and Business.* (In Greek)

Psarakis, S. and Panaretos, J. (1990). On an Approach to Model Evaluation and Selection, *XV Biometric Conference. Budapest, Hungary, July 1990.*

Xekalaki, E. and Katti, S. K. (1984). A Technique for Evaluating Forecasting Models. *Biometrical Journal*, 26(1), 173-184.

Received: January 1999, Revised: May 1999

17
CONVERGENCE RATE ESTIMATES IN FUNCTIONAL LIMIT THEOREMS

V. M. Zolotarev
Steklov Mathematical Institute
Russian Academy of Sciences

(To the honour of my Greek colleagues and friends)

1. Introduction. The Statement of the Problem

The following fact, sometimes called continuity theorem, is well known in probability theory (see Parthasarathy, 1980 or Billingsely, 1968).

Let us consider two Banach spaces $U = (\cdot, \|u\|_U)$ and $V = (\cdot, \|v\|_V)$ with norms of elements $\|u\|_U$ and $\|v\|_V$ respectively.

Let us also consider a function $f : U \to V$ and a sequence of random elements $\{u_n : n = 0, 1, 2, ...\}$, such that u_n weakly converges to u_0 as $n \to \infty$, that is

$$u_n \xrightarrow{w} u_0, \text{ as } n \to \infty. \tag{1}$$

By using the Lévy-Prokhorov distance $\pi(u, u')$ in the space U we can rewrite relation 0 in the form

$$\pi'_n = \pi(u_n, u_0) \to 0 \text{ as } n \to \infty.$$

Let us consider the sequence of random elements from V $\{v_n = f(u_n) : n = 0, 1, 2, ...\}$.

Continuity Theorem. Suppose that the function f is a continuous one. Then

$$\pi''_n = \pi(v_n, v_0) \to 0 \text{ if } \pi'_n \to 0 \text{ as } n \to \infty. \tag{2}$$

Obviously, it is quite reasonable to investigate the rate of convergence in 0 and, in particular, to estimate π''_n through π'_n in the traditional form

$$\pi''_n < \varphi(\pi'_n), \tag{3}$$

where $\varphi(x)$ is a nonnegative function on the halfline $x > 0$ with the property

$$\varphi(+0) = 0.$$

It is clear that we will not be able to obtain an estimate such as 0 only under the assumption of continuity of f. This assumption is too weak. Stronger assumptions are required such as the Lipschutz condition

$$\| f(u) - f(u') \|_V \leq A \| u - u' \|_U, \tag{4}$$

where A is a positive constant, or the Hölder condition

$$\| f(u) - f(u') \|_V \leq A^\alpha \| u - u' \|_U^\alpha, \text{ where } \alpha > 1. \tag{4'}$$

2. The main results.

Under these conditions we can prove the following theorems.

Theorem 1. Let the mapping f obey condition 0. Then the following inequality is true

$$\pi_n'' \leq A \pi_n'. \tag{5}$$

If f obeys condition (4') instead of (4), then

$$\pi_n'' \leq A^\alpha (\pi_n')^\alpha. \tag{5'}$$

Let us consider a special case of estimates (5) and (5') in application to a special case of the problem mentioned above. Namely, let us choose the spaces U and V as R^k, $k \geq 1$. We shall consider also the sequence of random elements as the sums

$$u_n = n^{-1/2} \sum (X_j : 1 \leq j \leq n),$$

where $\{X_j : j \geq 1\}$ are random vectors from R^k with independent identically distributed components with zero expectation, unit variance and finite third absolute moments, and u_0 is a random vector from R^k with independent and identically standard normal distributed components.

Sine by the central limit theorem we have

$$u_n \xrightarrow{w} u_0, \text{ as } n \to \infty,$$

by the continuity theorem for any continuous function f we have correspondingly

$$\pi_n'' = \pi(f(u_n), f(u_0)) \to 0, \text{ when } n \to \infty. \tag{6}$$

Using the method of metric distances we can essentially sharpen relation (6)

Theorem 2. Suppose that the function f satisfies condition (4). Then the following estimate is true

17. CONVERGENCE RATE ESTIMATES

$$\pi_n'' \leq Cn^{-1/8} \text{ for all } n \geq 1, \tag{7}$$

where C is an absolute constant.

Under condition (4') we obtain the estimate

$$\pi_n'' \leq An^{-\alpha/8}. \tag{7'}$$

3. Proof of Theorem 2

For the proof we shall need some lemmas.

Let us consider the ideal metric $\zeta_s(u,u')$ in the space R^k. It is well known (see, Billingsley, 1968) that, for the normalized sums for i.i.d. random vectors $X_1, X_2, ..., X_n$, we have

$$S_n = n^{-1/2} \sum (X_j : 1 \leq j \leq n).$$

Then, for a vector N having a standard normal distribution in R^n the following lemma holds.

Lemma 1

$$\zeta_s(S_n, N) \leq c.$$

Let us now consider random elements u_n and u_0 mentioned above and the ideal metrics $\zeta_s(u_n, u_0)$.

Lemma 1'. The following inequality is true

$$\zeta_s(u_n, u_0) \leq C_s n^{-(s/2-1)}, \tag{8}$$

where C_s is a positive constant depending on $s > 0$.

This inequality is a corollary of the main properties of the ideal metric ζ_s (see, Billingsley, 1968).

Lemma 2 provides a connection of the ideal metric ζ_s to the Lévy-Prokhorov metric π.

Lemma 2

$$\pi^{m+1}(u_n, u_0) \leq \zeta_m(u_n, u_0). \tag{9}$$

This inequality was proved by the Bulgarian mathematician G. Yamukov (1977).

Firstly we shall prove inequality 0. More precisely, under assumption 0, inequality

$$\pi(f(u), f(u')) \leq \pi(u, u'). \tag{10}$$

It is known that the Lévy-Prokhorov metric $\pi(u, u')$ is the minimal metric for the Ky-Fan metric

$$K(u, u') = \inf\{\varepsilon > 0 : P(\|u - u'\|_U < \varepsilon) \leq \varepsilon\},$$

that is

$$\pi(u, u') = \inf K(u, u'),$$

where inf is considered over all common distributions of u and u'. Hence, using condition 0, we have

$$\{\|f(u) - f(u')\|_V < \varepsilon\} \subseteq \{A\|u - u'\|_U < \varepsilon\}$$

and, as a corollary,

$$K(f(u), f(u')) \leq \inf\{\varepsilon > 0 : P(A\|u - u'\|_U < \varepsilon) \leq \varepsilon\} = K(u, u').$$

Replacing variable ε by $A\eta$, we obtain

$$K(f(u), f(u')) \leq \inf\{A\eta : P(\|u - u'\|_U < \eta) \leq \eta\},$$

that is

$$K(f(u), f(u')) \leq AK(u, u').$$

By minimizing both sides of the above inequality, we obtain

$$\pi(f(u), f(u')) \leq A\pi(u, u'), \tag{11}$$

which leads to (7).

Inequality 0 is a simple corollary of this inequality. Obviously, inequality 0 is the basis of the method of proof, therefore we will include this fact in a number of lemmas:

Lemma 3. For any function f on U, satisfying condition 0, and an arbitrary pair of random elements u, u' from U, inequality 0 is true.

To prove lemma 3, let us consider the Ky-Fan distance between $v = f(u)$ and $v' = f(u')$:

$$K(u, u') = \inf\{\varepsilon > 0 : P(\|v - v'\|_V > \varepsilon) \leq \varepsilon\},$$

that is

$$K(v, v') = \inf\{\varepsilon > 0 : P(\|f(u) - f(u')\|_U > \varepsilon) \leq \varepsilon\}.$$

By using inequality 0, we can write

$$P(\|f(u) - f(u')\|_U > \varepsilon) \leq P(A\|u - u'\|_U > \varepsilon),$$

and then replace variable ε with $A\varepsilon$.

To demonstrate the method of proof of these statements we will consider in detail the proof of Theorem 2. To prove Theorem 2, we will need a number of additional lemmas. We will have to deal with two probability metrics: The Lévy-Prokhorov metric $\pi(u, u')$ in $U = R^k$ and the ideal metric $\varsigma_s(u, u')$ of order $s > 0$.

17. CONVERGENCE RATE ESTIMATES

General information on probability metrics can be found in Parthasarathy, (1980) or in Zolotarev, (1983). In particular, we note the following properties of metric π and ς_s in R^k.

$$\pi(u, u') = \inf K(u, u'),$$

where $K(u, u')$ is called Ky-Fan metric

$$K(u, u') = \inf\{\varepsilon > 0 : P(\|u - u'\|_U > \varepsilon) \leq \varepsilon\},$$

and inf is taken over all common distributions of random pairs (u, u') with fixed marginal distributions. The talented Bulgarian mathematician G. I. Yamukov was able to find the following connection between metrics π and ς_s:

Lemma 4. (Theorem of G. I. Yamukov, 1977)

For any random elements u, u' from $U = R^k$ and an arbitrary positive index s,

$$\pi^{1+s}(u, u') \leq C'_s \varsigma_s(u, u'),$$

where C'_s is a positive constant depending only on s. In particular, $C_1 = 2$, $C_2 = 24$.

Now let us collect all inequalities, which help to estimate $\pi(f(u_n), f(u_0))$. Namely,

$$\pi(f(u), f(u')) \leq A\pi(u, u'), \tag{7}$$

$$\varsigma_s(u_n, u_0) \leq C_s n^{-(s/2 - 1)}, \quad s > 0. \tag{8s}$$

$$\pi(u, u') \leq C'_m \varsigma_m(u, u'), \tag{9}$$

As a corollary of these inequalities we can obtain the following estimates:

$$\pi^4(v_n, v_0) \leq \pi^4(f(u_n), f(u_0)) \leq A^4 \pi^4(u_n, u_0).$$

Since

$$\pi^4(u_n, u_0) \leq C_3 n^{-1/2},$$

we have

$$\pi(f(u_n), f(u_0)) \leq C n^{-1/8}.$$

Analogously, if we use (4') instead of (4), we obtain

$$\pi(f(u_n), f(u_0)) \leq C_0 n^{-\alpha/8}, \quad \alpha \geq 1.$$

It is not difficult to see that the main part of the proof of Theorem 2 is inequality (7). So, there is a reason to find a more general condition than (4). The following theorem gives an answer to this question.

Theorem 3. Suppose that the function f satisfies condition

$$\| f(u) - f(u') \|_V \leq f(\| u - u' \|_U; \| u' \|_U),$$

where $f(x, y)$ is a nonnegative function, which is monotone on both variables x and y with

$$f(+0, y) = 0 \text{ for any } y.$$

Let us denote $\pi_1 = \pi(u, u')$ and $\pi_2 = \pi(f(u), f(u'))$. Then,

$$\pi_2 \leq f(\pi_1, 1 - \pi_1). \tag{10}$$

It is clear that inequality (7) is a special case of (10).

A proof of Theorem 3 will be given in a separate paper.

4. Corollary (An illustration of the application of Theorem 2)

In their paper, Professors Goetze, Prokhorov, and Ulyanov (1996) prepared the analytical basis to solve the problem of estimation of the difference between the distributions of two random polynomials. They do it in a traditional way with the help of the method of characteristic functions.

In the sequel, we solve a similar problem but with the use of Theorem 2, based upon the method of metric distances, which was just demonstrated in section 3.

For the sake of simplicity, we consider a special case of the general problem. Namely, let

$$\mathbf{X} = (X_1, X_2, \ldots, X_n, \ldots), \quad \tilde{\mathbf{X}} = (\tilde{X}_1, \tilde{X}_2, \ldots, \tilde{X}_n, \ldots)$$

be two sequences of i. i. d. r. v.'s with zero mean, unit variance and finite third absolute moment. Now let us form using $\mathbf{X}$ and $\tilde{\mathbf{X}}$ two new sequences of random variables $P(\mathbf{X})$ and $\tilde{P}(\tilde{\mathbf{X}})$:

$$\{P_m = a_1 X_1 X_2 + a_2 X_3 X_4 + \ldots + a_{m/2-1} X_{m/2-1} X_m\}, \quad m = 1, 2, \ldots,$$

and a similar sequence of random polynomials

$$\{\tilde{P}_m = a_1 \tilde{X}_1 \tilde{X}_2 + a_2 \tilde{X}_3 \tilde{X}_4 + \ldots + a_{m/2-1} \tilde{X}_{m/2-1} \tilde{X}_m\}, \quad m = 1, 2, \ldots.$$

If we assume that symmetric coefficients in the polynomials are equal, then the polynomials will be symmetric and homogeneous of order 2.

17. CONVERGENCE RATE ESTIMATES

For such polynomials,

$$\frac{1}{n}P_n(\mathbf{X}) = P_n\left(\frac{1}{\sqrt{n}}\mathbf{X}\right) \text{ and } \frac{1}{n}P_n(\widetilde{\mathbf{X}}) = P_n\left(\frac{1}{\sqrt{n}}\widetilde{\mathbf{X}}\right),$$

because

$$S_n = \frac{1}{\sqrt{n}}\mathbf{X}_n \xrightarrow{w} N, \text{ as } n \to \infty,$$

where $\mathbf{X} = (X_1, X_2, \ldots, X_n, \ldots)$. By the continuity theorem, we can state that

$$P_n = P(S_n) \to P(N),$$

where N is a normal random vector of R^n.

To use theorem 2, firstly we have to define a norm for random polynomials. We can do it as follows:

$$\|P_m(\mathbf{X})\| = \left(E|P_m(\mathbf{X})|^2\right)^{1/2}.$$

In this case, condition (4) has the following form

$$\|\text{grad}P_m(\mathbf{X})\| = \left(\sum_j a_j^2\right)^{1/2} \leq 1. \tag{4''}$$

After that, the corollary of Theorem 2 will be as follows:

Corollary. Under the above-mentioned conditions and under assumption (4''), we have

$$\pi\left(P_n(S_n), \widetilde{P}_n(\widetilde{S}_n)\right) \leq cn^{-1/8},$$

where c is a positive absolute constant.

Acknowledgments

I thank Victor Panaretos, who was the first attentive reader of my chapter and made some useful comments. I also thank my student, Maria Vlushina, for her help in the preparation of this paper.

5. References

Billingsley, P. (1968). Convergence of Probability Measures. Wiley, New York.

Goetze, F, Prokhorov, Yu and Ulyanov, V. (1996). Estimations of the Characteristic Functions for the Polynomials, Depending on Asymptotically Normal Random Variables, published in the preprints of Bielifeld University (1996).

Parthasarathy, K. R. (1980). Introduction to Probability and Measure.

Yamukov, G. L. (1977). Estimations for Generalized Dudley Metrics in the Spaces R^k, $k>1$. *Probability Theory and Applications*, v. 22, No. 3, 590-595.

Zolotarev, V. M. (1983). Probability Metrics. *Probability Theory and Applications*, v. 28, No. 2.

Received: April 2002, Revised: May 2002

Author Index

A

Abbott, A., 53, 67
Abd-El-Hakim, N.S., 81, 92
Abramovitz, M., 176, 178
Abramowitz, M., 192, 193, 201
Aitkin, M., 83, 86, 92, 117, 127
Akaike, H., 124, 127
Alanko, T., 81, 92
Alberts, B., 50, 67,
Alexander, C.H., 100, 102
Al-Husainni E.K., 81, 92
Alsmeyer, G., 167, 178
Alves, M.I.F., 156
Anderson, C. W., 2, 11,
Anderson, D., 86, 92
Andersen P.K., 27, 28
Andrews, D.F., 81, 92
Angrist, J.D., 25, 28, 53, 67
Arminger, G., 117, 127
Aromaa, A., 60, 69
Aurelian, L., 60, 67

B

Banfield, D.J., 83, 92
Banks, L., 60, 71
Barão, M.I., 141, 156
Barlow, R.E., 164, 165, 166
Barndorff-Nielsen, O.E., 22, 28, 81, 92
Barnett, V. 1, 3, 4, 7, 9, 11
Bartholomew, D.J., 13, 14, 17, 18, 19, 87, 92, 117, 123, 127
Basford, K., 83, 91, 92, 94
Beirlant, J., 141, 144, 146, 153, 154, 157
Bender, P., 117, 127
Berger, J.O., 22, 28
Berke, O., 182, 187
Berkson, J., 62, 67
Bernardo, J.M., 22, 28
Berred, M., 148, 157
Billingsley, P., 203, 205, 209
Bingham, N.H., 172, 173, 178
Blau, P. M., 64, 67
Blossfeld, H-P., 27, 28
Bock, R. D., 117, 127
Böhning, D., 79, 91, 92, 93
Bolt, H.M., 185, 187
Boos, D.D., 149, 157
Borgan, O., 27, 28
Borkowf, C. B., 103, 116
Bose, S., 88, 93
Bound, J. A., 62, 68
Bradford Hill, A., 24, 28

Bray, D., 50, 67
Brazauskas, V., 153, 157
Breslow, N., 56, 62, 67
Breuer, J., 60, 71
Brody, H., 50, 70
Broniatowski, M., 146, 157
Brooks, R.J., 84, 93
Brooks, S.P., 81, 93
Bross, I. D. J., 63, 65, 67
Brown, R. L., 190, 201
Buck, C., 60, 67
Buck, R.J., 27, 28
Brychkov, Y.A., 136, 140

C

Cann, C. I., 66, 69
Cannistra, S. A., 60, 67
Cao, G., 84, 93
Caperaa, P., 156, 157
Carmelli, D., 63, 67
Caroll, R.J., 90, 93, 103, 116
Carpenter, K. J., 50, 67
Chaubey, Y. P., 89, 94
Chen, S.X., 86, 93
Cleland, J., 125, 127
Clogg, C.C., 87, 91, 94, 95
Cokburn, I., 85, 95
Coles, S.G., 141, 142, 157
Colwell, R. R., 50, 67
Conforti, P. M., 54, 69
Cook, D., 62, 67
Cooper, R. C., 54, 69
Copas, J. B., 57, 67
Cornfield, J., 56, 60, 62, 64, 67, 71
Cowell, R.G., 24, 28
Cox, D.R., 20, 22, 24, 27, 28, 58, 68
Csörgő, S., 147, 157

D

Dacey, M. F., 81, 93
Dacorogna, M.M., 149, 159
Dalal, S. R., 88, 93
Danielsson, J., 141, 145, 155, 157
Darby, S. C., 59, 68
David, H. A., 2, 4, 11
Davies, G., 6, 11
Davis, H. J., 60, 67
Davis, J., 59, 71
Davis, R., 144, 149, 157
Davison, A.C., 141, 150, 157
Dawid, A.P., 24, 28
Day, N. E., 56, 62, 67

Dean, C.B., 85, 93
Deheuvels, P., 145, 147, 149, 157
Dekkers, A.L.M., 143, 144, 145, 154, 157
Demetrio, C. G. B., 85, 93
Dempster, A. P., 183, 186
Desrosières, A., 53, 68
Devroye, L., 89, 93
De Haan, L., 143, 144, 145, 149, 154, 155, 156, 157, 158
De Ronde, J., 156, 157
De Vries, C.G., 141, 145, 155, 157
De Vylder, F., 175, 178
De Wolf, P.P., 148, 149, 158
Dickersin, K., 58, 68
Diebolt, J., 92, 93
Dierckx, G., 146, 157
Dijkstra, T. K., 56, 68
Dixon, M.J., 142, 157
Doering, C. R., 60, 69
Dolby, G.R., 14, 19
Doll, R., 59, 60, 61, 62, 64, 68
Doraiswami, K. R., 60, 71
Downes, S., 59, 69
Draisma, G., 155, 157
Drees, H., 150, 152, 154, 157, 158
Dubos, R., 50, 68
Duffy, J.C., 81, 92
DuMouchel, W.H., 142, 158
Duncan, O. D., 65, 67
Durbin, J., 190, 201
Dynkin, E.B., 172, 178

E

Eaton, M.L., 162, 166
Eberlein, E., 81, 93
Edler, L., 185, 186
Efron, B., 29
Ehrenberg, A. S. C., 62, 68
Einmahl, J.H.J., 144, 145, 157
Embrechts, P., 141, 142, 143, 144, 156, 158
Emrich, K., 179, 185, 186
Escobar, M., 88, 93
Evans, A. S., 60, 62, 68
Evans, H. J., 59, 68
Evans, J. M., 190, 201
Evans, R. J., 50, 68
Everitt, B. S., 83, 86, 91, 93

F

Fang, H. B., 130, 139
Fang, K. T., 129, 130, 131, 139
Fayers, P.M., 126, 127
Feller, W., 167, 178
Feuerverger, A., 112, 116, 148, 158

Fialkow, S., 50, 68, 70
Fildes, R., 22, 28
Finlay, B. B., 50, 68
Fisch, R.D., 179, 187
Fisher, R. A., 62, 68, 98, 102, 141, 158
Fougeres, A.L., 156, 157
Freedman, D., 45, 50, 53, 55, 56, 57, 58, 67, 68, 69
Friedman, M., 65, 68

G

Gail, M. H., 64, 68, 103, 116
Gagnon, F., 60, 68
Galbraith, J., 123, 127
Gamble, J. F., 59, 68
Gani, J., 72, 74, 77
Gao, F., 59, 71
Gardiol, D., 60, 71
Gardner, M. J., 59, 68
Gauss, C. F., 51, 69
Gavarret, J., 46, 69
Geluk, J.L., 155, 158
Gilberg, F., 185, 186
Gill, R.D., 27, 28, 103, 116
Gleser, L. J., 81, 93
Glymour, 24, 28
Goegebeur, Y., 146, 157
Goetze, F, 208, 210
Gold, L. S., 56, 68
Goldie, C.M., 172, 173, 178
Goldstein, H., 89, 93
Goldthorpe, J. H., 53, 69
Golka, K., 185, 187
Gomes, M. I., 2, 6, 11, 144, 149, 156, 158
Goovaerts, M.J., 175, 178
Gouriéroux, C., 188, 201
Goutis, K., 91, 93
Gradshteyn, I. S., 131, 133, 134, 136, 138, 139
Graham, E. A., 60, 71
Greenland, S., 58, 66, 69, 70
Greenwood, M., 81, 93
Grego, J., 87, 94
Grimshaw, A., 150, 158
Grimson, R., 83, 95
Groeneboom, P., 148, 149, 158
Groves, R.M., 101, 102
Guillou, A., 154, 158
Gumbel, E.J., 142, 158
Gupta, S., 91, 93
Gupta, R., 81, 95
Gurland, J., 81, 95

AUTHOR INDEX

H

Haenszel, W., 56, 67
Haeusler, E., 144, 158
Haezendonck, J., 175, 178
Hakama, M., 60, 69
Hall, A. J., 59, 69
Hall, P., 90, 93, 148, 149, 154, 158
Hall, W. J., 88, 93
Hamerle, A., 27, 28
Hammond, E. C., 56, 67
Hand, D.J., 91, 93, 126, 127
Hansen, M.H., 97, 102
Harwood, C., 60, 71
Hasofer, A.M., 156, 158
Hasselblad, V., 82, 92, 93
Hebert, J., 81, 93
Heckman, J. J., 53, 69
Hedenfalk., 38
Heffron, F., 68
Hering, F., 186, 187
Hill, A. B., 60, 61, 62, 68
Hill, B.M., 144, 158
Hinde, J., 85, 86, 92, 93
Hinderer, K., 170, 178
Hodges, J. L., 53, 69
Holland, P., 53, 69
Hosking, J.R.M., 142, 150, 156, 158
Hosmer, D., 82, 93
Howard-Jones, N., 50, 69
Hsing, T., 144, 158
Huang, W.T., 91, 93
Huang, X., 156, 158
Humphreys, P., 53, 69
Huq, N. M., 125, 127
Hurwitz, W.N., 97, 102

I

Imbens, G. W., 53, 67
Irle, A., 72, 74, 77
Irwin, J. O., 81, 84, 93
Iyengar, S., 129, 133, 139
Izawa, T., 195, 202

J

Jackson, L. A., 54, 69
Jansen, D.W., 145, 157
Janssen, J., 27, 28
Jansen, R.C., 179, 187
Jeffreys, H., 22, 28
Jensen, A., 97, 102
Jewell, N., 81, 93
John, S. M., 60, 69
Johnson, N.L., 78, 94, 103, 114, 116
Jöreskog, K. G., 117, 127

Jorgensen, B., 81, 94

K

Kalita, A., 60, 71
Kanarek, M. S., 54, 69
Kao, C.-H., 179, 180, 186, 187
Kaprio, J., 63, 69
Karlin, B., 88, 94
Karlin, S., 167, 178
Karlis, D., 78, 81, 83, 94, 185, 186, 187
Katti, S. K., 188, 196, 202
Kaufmann, E., 154, 157
Kaur, A., 9, 11
Keiding, N., 27, 28
Kellerer, H.G., 163, 166
Keller, U., 81, 93
Kemp, A.W., 78, 94
Kendall, M. G., 81, 94, 96, 101, 102, 190, 202
Kent, J., 81, 92
Keohane, R.O., 24, 28
Kiaer, A.W., 97, 102
Kibble, W. F., 192, 202
King, G., 24, 28
Kinlen, L. J., 60, 69
Kish, L., 96, 99, 100, 101, 102
Klüppelberg, C., 141, 142, 143, 144, 158
Knekt, P., 60, 69
Knott, M., 14, 18, 19, 117, 118, 123, 126, 127
Kogon, S.M., 148, 158
Koskenvuo, M., 63, 69
Kotlarski, I., 190, 191, 202
Kotz, S., 78, 94, 103, 114, 116, 129, 130, 131, 132, 133, 134, 139, 140
Kovacs, J., 182, 187
Kratz, M., 146, 159
Krishnaiah, P.R., 163, 164, 166
Krishnaji, N., 90, 94
Krishnan, T., 92, 94
Krueger, A.B., 25, 28
Kruskal, W.H., 97, 102
Kusters, U., 117, 127

L

Laird, N., 88, 94
Laird, N. M., 183, 186
Lamperti, J., 172, 178
Lancaster, T., 27, 28
Lang, J. M., 66, 69
Lauritzen, S.L., 24, 28
Lawless, J., 85, 93, 94
Lawley, D. N., 117, 127
Le, N., 85, 95

Lee, S.-Y., 18, 19, 85, 94, 117, 127
Legendre, A. M., 51, 69
Legler, J., 117, 127
Lehmann, E. L., 53, 69, 162, 164, 165, 166
Lehtinen, M., 60, 69
Leigh, I. M., 60, 71
Leinikki, P., 60, 69
Leroux, B., 90, 94
Lewis, J., 50, 67
Lewis, T., 3, 4, 11, 88, 94
Li, H. G., 57, 67
Lieberson, S., 53, 69
Lieblein, J., 2, 11
Lin, T. H., 56, 68
Linardis A., 196, 202
Lindsay, B., 79, 87, 91, 92, 94
Lindsay, B.G., 91, 95
Liu, M.C., 90, 94
Llopis, A., 60, 67
Lloyd, E. H., 4, 11
Lilienfeld, A. M., 56, 67
Liu, T. C., 53, 69
Lombard, H. L., 60, 69
Lopuhaa, H.P., 148, 149, 158
Louis, P., 45, 69
Lovejoy, W., 99, 102
Lucas, R. E. Jr., 53, 69
Lwin, 88, 94

M

Madow, W.G., 97, 102
Makridakis, S., 22, 28
Mallet, A., 85, 94
Mallows, C. L., 81, 92
Makov, U.E., 79, 91, 95
Manski, C. F., 53, 70
Mantovani, F., 60, 71
Maraun, M.D., 16, 19
Marcus, R. L., 60, 67
Marichev, O.I., 136, 140
Maritz, J. L., 88, 94
Márkus, L., 182, 187
Marohn, F., 156, 159
Martins, M.J., 144, 149, 158
Mason, D., 147, 157
Matlashewski, G., 60, 71
Matths, G., 146, 157
Maxwell. A. E., 117, 127
Mayer, K.U., 27, 28
McCullagh, P., 117, 127
McIntyre, G. A., 7, 11
McKim, V., 53, 70
McLachlan, G., 82, 83, 92, 94
McLachlan, J.A., 79, 91, 92, 94
McMillan, N., 59, 71

McNeil, A.J., 141,159
Mejza, S., 186, 187
Mekalanos, J. J., 50, 70
Melchinger, A.E., 186, 187
Merette, C., 83, 93
Miettinen, A., 60, 69
Mikosch, T., 141, 142, 143, 144, 158, 159
Mill, J. S., 45, 70
Miller, J. F., 50, 70
Mitchell, T.J., 27, 28
Mittnik, S., 148, 159
Monfort A., 188, 201
Mooney, H.W., 99, 102
Moore, K.L., 9, 11
Morgan, B.J.T., 81, 93
Morris, M.D., 27, 28
Mosteller, F., 97, 102
Moustaki, I., 18, 19, 117, 118, 119, 123, 126, 127
Mudholkar, G., 89, 94
Müller, F. H., 60, 70
Muller, U.A., 149, 159
Murchio, J. C., 54, 69
Muthen, B., 117, 127

N

Nadarajah, S., 129
Nagaraja, H.N., 148, 159
Nájera, E., 60, 67
Navidi, W., 55, 68
Nelder, J.A., 85, 94
Nelder, J., 117, 127
Neves, M., 149, 158
Neyman, J., 23, 28, 53, 70, 97, 102
Ní Bhrolcháin, M., 53, 70
Nicolaides-Bouman, A., 64, 71
Niloff, J. M., 60, 67
Ng, K. W., 129, 130, 131, 139

O

Oakes, D., 27, 28
O'Muircheartaigh, C., 97, 102
Ord, J. K., 2, 12, 190, 202
Ostrovskii, I., 129, 133, 134, 140
Ottenbacher, K. J., 58, 70

P

Paavonen, J., 60, 69
Pack, S.E., 81, 93
Page, W. F., 63, 67
Panaretos, J., 90, 94, 95, 141, 188, 202
Paneth, N., 50, 70
Paolella, M.S., 148, 159
Pardoel, V. P., 56, 71

AUTHOR INDEX

Parthasarathy, K. R., 203, 207, 210
Pasteur, L., 70
Patil, G. P., 9, 11
Pearl, J., 24, 28, 53, 58, 69, 70
Pearson, E.S., 23, 28
Peel, D., 79, 91, 92, 94
Peng, L., 145, 149, 150, 155, 157, 158, 159
Pereira, T.T., 145, 149, 155, 157
Perneger, T. V., 58, 70
Petitti, D., 56, 68
Peto, R., 60, 69
Philippe, A., 89, 94
Pickands, J., 143, 159
Pictet, O.V., 149, 159
Pisani, R., 58, 68
Poon, W.-Y., 18, 19, 117, 127
Pope, C. A., 59, 70
Porter, T.M., 96, 102
Rothman, K. J.
Powell, C. A., 59, 69
Prokhorov, Yu, 208, 210
Proschan, F., 164, 165, 166
Prudnikov, A.P., 136, 140
Psarakis, S., 188, 202
Purkayastha, S., 9, 11
Purves, R., 58, 68
Puterman, M., 85, 95

Q

Quetelet, A., 70

R

Rachev, S.T., 81, 94, 148, 159, 163, 166
Raff, M., 50, 67
Raftery, A.E., 83, 92
Rackow, P., 99, 102
Ransom, M. R., 59, 70
Rao, C.R., 90, 94, 161, 162, 166
Rao, M. B., 161, 163, 164, 166
Raufman, J. P., 50, 70
Redner R., 91, 95
Reiss, R.D., 152, 158
Renault E., 188, 201
Resnick, S.I., 141, 144, 146, 149, 151, 152, 154, 157, 158, 159
Ridout, M.S., 81, 93
Rip, M., 50, 70
Robert, C., 92, 93
Roberts, K., 50, 67
Robins, J. M., 57, 58, 69, 70
Robinson, W. S., 49, 70
Rohwer, G., 27, 28
Røjel, J., 60, 70
Rootzén, H., 141, 150, 159

Rosen, O., 149, 159
Rosenbaum, P.R., 25, 28
Rosenberg, C. E., 50, 70
Ross, S.M., 167, 168, 171, 178
Rothman, K. J., 58, 66, 69, 70
Rotnitzky, A., 57, 70
Rubin, D. B., 53, 67, 70, 183, 186
Rudas, T., 91, 95
Ruschendorf, L., 163, 166
Ryan, L., 117, 127
Ryzhik, I. M., 131, 133, 134, 136, 138, 139

S

Sachs J., 27, 28
Sacks, J., 27, 28, 59, 71
Sammel, M., 117, 127
Samorodnitsky, G., 141, 158
Sarhan, A. E., 10, 11
Scallan, A. J., 81, 95
Scarf, P.A., 142, 159
Scharfstein, D. O., 57, 70
Scheines, 24, 28
Schön, C.C., 186, 187
Schroff, P. D., 60, 71
Schumann, B., 60, 67
Schwartz, J., 59, 70
Schweizer, B., 112, 116
Sclove, S., 124, 128
Segers, J., 144, 148, 156, 159
Selinski, S., 185, 187
Semmelweiss, I., 50, 70
Sengupta, A., 81, 94
Serfling, R., 153, 157
Seshadri V., 81, 94
Shaked, M., 80, 95
Shanbhag, D.N., 161, 162, 166
Shephard, N., 22, 28
Shimkin, M. B., 56, 67
Sibuya, M., 91, 95
Silverman, B.W., 86, 95
Simon, H. A., 81, 95
Simonoff, J.S., 86, 95
Sinha, A.K., 156, 157
Sinha, B. K., 9, 11
Sinha, R. K., 9, 11
Smith, A.F.M., 79, 91, 95
Smith, R.L., 141,142, 150, 157, 159
Snee, M. P., 59, 69
Snow, J., 45, 70
Sörbom, D., 117, 127
Sorensen, M., 81, 92
Spiegelhalter, D.J., 24, 28
Spirtes, 24, 28
Stărică, C., 144, 151, 152, 154, 159
Steele, F., 123, 127

Stegun, I.A., 176, 178, 192, 193, 201
Stephens, D.A., 179, 187
Stigler, S. M., 53, 70, 96, 102
Stokes, S. L., 9, 11
Stolley, P., 63, 70
Storey, A., 60, 71
Streit, F., 129, 133, 134, 140
Stuart, A., 2, 12, 190, 202
Styer, P., 59, 71
Subramanyam, K., 163, 164, 166
Symons, M., 83, 95

T

Taillie, C., 9, 11
Tajvidi, N., 141, 150, 159
Taubes, G., 56, 71
Tawn, J.A., 141, 156, 157
Taylor, H.M., 167, 178
Teasdale, R. D., 186, 187
Teppo, L., 60, 69
Terrell, J. D., 59, 69
Terris, M., 50, 60, 67, 71
Teugels, J.L., 141, 144, 146, 153, 154, 156, 157, 158, 159, 167, 172, 173, 178
Thomas, M., 60, 71
Tippet, L.H.C., 141, 158
Titterington, D.M., 79, 91, 95
Tong, Y. L., 129, 133, 139
Tripathi, R., 81, 95
Tsourti, Z., 141
Tukey, J. W., 2, 12
Turner, S., 53, 70
Tzamourani, P., 123, 127

U

Ulyanov, V., 208, 210
Urfer, W., 179, 182, 185, 186, 187
Utz, H.F., 186, 187

V

Vandenbroucke, J. P., 56, 71
Van Ryzin, J., 86, 95
Verba, S., 24, 28
Venn, J., 101, 102
Vinten-Johansen, P., 50, 70
Von Mises, R., 98, 102

Vynckier, P., 141, 144, 146, 153, 154, 157

W

Wald, A., 23, 28
Wald, N., 64, 71
Walker H., 91, 95
Wallis, J.R., 142, 150, 158
Wang, M.C., 86, 95
Wang, P., 85, 95
Wang, Z., 156, 158
Watson, J. D., 50, 67
Wedderburn, R., 117, 127
Weissman, I., 149, 159
Welch, W.J., 27, 28
Wermuth, N., 24, 28
West, M., 84, 88, 93
Whitmore G.A., 81, 94
Williams, D.B., 148, 158
Willmot, G.E., 85, 93
Wilson, T., 83, 92
Winkelstein, W., 50, 71
Wolfe, J.H., 92, 95
Wolff E. F., 112, 116
Wong, S.T., 97, 102
Wood, E.F., 142, 158
Wynder, E. L., 56, 60, 67, 71
Wynn, H.P., 27, 28

X

Xekalaki, E., 78, 81, 83, 84, 90, 94, 95, 185, 186, 187, 188, 196, 202

Y

Yamukov, G. L., 205, 207, 210
Yuan, Y., 83, 95
Yule, G., 81, 93
Yule, G. U., 45, 51, 71, 81, 95
Yun, S., 144, 159

Z

Zeisel, H., 56, 68
Zeng, Z.-B., 179, 180, 186, 187
Zolotarev, V. M., 203, 207, 210

Subject Index

A

Age,
    Golden, of statistics, 42
    Normalized, 174-175
    Of the renewal process, 167

Algorithms,
    Automatic, 35
    ECM, 179, 186
    EM, 32, 33, 82, 89, 91, 92, 123, 183, 185, 186
    Iterative, 154
    Monte-Carlo, 179
    Prediction, 37

Analysis
    Cluster, 83
    Discriminant, 82-83
    Ecological, 48
    Extreme-value, 144, 146, 148, 156
    Factor, 16, 86-88, 124
    Louis's, 46
    Of Variance 35, 42, 84
    Path, 24
    Preliminary data, 56, 57
    Survival, 32
    Sensitivity, 26-27

Approach
    Bayesian, 22, 23, 88
    Bootstrap, 155, 156
    Box-Cox transformation, 33
    Distribution-Free to outliers, 5-7
    EM algorithmic, 91
    Maximum Domain, 142
    Modelling, 45, 66, 142-143
    Neyman's, 31
    Non-model-based, 2
    Non-parametric, 142
    Parametric, 142
    Personalistic, 22
    QQ-plot, 146
    Regression, 154
    Response function, 117
    Robust inference, 3
    Selection, 188-202
    Semi-parametric, 142
    Variable, 117
    Yule's, 53

B

Bayes,
    Empirical, 22, 32, 33, 40, 41, 42, 43
    Estimation, 88
    Frequentist Compromise 34
    Methodologies, 88
    Rule, 40
    Subjective formulation, 31
    Theorem, 87

Bayesian,
    Applications, 41
    Approaches, 22
    Empirical, 86
    Formulations, 22
    Justification, 41
    Objective conclusions, 35
    Robustness, 88

Bias,
    Asymptotic square, 145
    Of extreme-value index estimators, 156
    Of the Hill estimator, 145
    Of the standard estimators, 146
    "recall", 46
    "selection", 47, 57
    Variance, trade-off, 150, 152

Biometry, 20

Bootstrap, 22, 33, 36, 43, 91
    Approach, 155
    Double, 155
    Methodology, 149, 156
    Replication, 155
    Resamples, 149
    Samples, 155

C

Calibration, 26

Causal
    Association, 52
    Inferences, 45, 53, 65, 66
    Interpretation, 24

Chi-Square
    Mixing distribution, 81
    Pearson's, 30
    Value, 126
    Variable, 190

Cholera,
    Snow on, 47

Coefficients, 51, 52, 53, 55, 56, 57, 189
    Correlation, 89, 192, 196
    Estimated factor, 126

Moment, 17
Partial regression, 181
Regression models, random, 85
Regression, 23, 24, 99, 189
Symmetric, 208
Cohort study, 62
Cointegration, 22
Comparison,
  Graphical, 152
  Multinational, 99
  Periodic, 99
Confounding, 46, 55
Condition,
  Bradford Hill's, 25
  Lipschutz, 204
  Hölder, 204
Consistency,
  Asymptotic, 148
  Strong, 144, 145, 148
  Testing of, 21
  Weak, 145, 146, 148, 149
Contaminants, 3, 13
Convex,
  Combination, 144, 162
  Generalized combination, 162
  Sets of multivariate distributions, 161-166
Correlation,
  Coefficient, 89, 192, 196
  Galton, 30
  Sample, 197
  Weak, 59

D

Data,
  Bell shaped, 86
  Categorical, Adequate Treatment of, 17-18
  Crop-Yield, 196-197
  Empirical, 96
  Genetic, 179
  J-shaped, 86
  Logarithmic-transformed, 147
  Mapping, 181
  Mining, 37, 43
  Missing, 82, 91, 182, 183, 184
  Modelling, 80-82
  Observed, 183
  Panel, 27
  Survey, 117
  Survival, 27
  Training, 82

Tumor, 40
Univariate, 3
Yield, 189
Decision theory,
  Statistical, 22-23
  Wald's, 31
Density,
  Bivariate, 173
  Conditional, 184
  Contours of Fréchet-Type Elliptical, 132
  Contours of Gumbel-Type Elliptical, 137
  Contours of Weibull-type, 132
  Discrete, 86
  Function, 183
  Generator, 129
  Joint, 130, 131-132, 134-135, 137-138, 173
  Joint function, 104
  Kernel, estimation, 85, 86
  Limiting, 173
  Marginal, 87
  Marginal, function, 84
  Mixing, 87
  Normal, 182
  Of a Weibull-type distribution, 129
  Prior, 22
  Probability, function, 2, 78, 79, 81, 82, 85, 103, 192, 193, 195, 196, 201
  Standard normal, 121
Dependent,
  Distributions on the unit score, 103-116
  Positive quadrant, 162
  Random variables, 189, 190, 193
Design,
  α, 186
  Case-control, 59
  Effects, 98, 99
  Experimental, 36, 42, 97
  Multipopulation, 99
  Probability, 99
  Research, 62
  Robust, 99
  Sample, 99
  Sampling, 9
  Study, 27, 45, 54, 62
Distance,
  Average standardized, 191
  Ky-Fan, 206
  Lévy-Prokhorov, 203
  Metric, method of, 204, 208
  Minimum Hellinger, 186

SUBJECT INDEX

Distribution,
    Standardized, 190
    Arising out of Methods of Ascertainment, 90
    α-stable, 148
    Binomial, 90
    Bivariate, 103
    Bivariate elliptical, 130
    Bernoulli, 119
    Beta, 81
    Beta type II, 195
    Cauchy, 31, 154, 155
    Complex, 98
    Conditional, 14, 15, 16, 18, 87, 90, 183, 184
    Continuous, 90
    Convex, 161
    Correlated Gamma Ratio, 188-202, 195, 196
    Exponential, 154, 170, 175
    Extreme point, 162
    F, 189, 191, 193, 195
    Finite step, 87
    Fixed marginal, 207
    Fréchet, Type II, 130, 134-137
    Gamma, 81
    Gamma mixing, 89
    Generalized extreme value, 141, 150, 154, 155
    Generalized Pareto, 150, 155
    Generic, 171, 172
    Gumbel, 3, 4, 5, 130
    Gumbel-Type elliptical, 137-139
    Inverted pyramid, 108
    Joint, 173, 174
    Joint, of manifest variables, 120, 121
    Kibble's Bivariate Gamma, 192, 193, 195
    Kotz-Type elliptical, 131-134
    Layered square tray, 104
    Layered tray, 116
    Limit, of uncorrelated but dependent, on the unit square, 103-116
    Logistic, 89
    Mixing, 80
    Multinomial, 120
    Multiple-level square tray, 103
    Multivariate elliptical, 129
    Multivariate, n-dimensional, 129
    Normal, 120, 129
    Multivariate, on convex sets, 161-166
    Negative binomial, 81, 89
    Normal, 84, 87, 89
    N-variate, 161
    Penultimate, 3, 5
    Poisson, 78, 81, 85, 86, 89, 98
    Positive Quadrant Dependent, 162-164
    Prior, 88
    Probability, 100
    Pyramid, 105-107
    Pyramid-type square, 112, 115
    Reserve-pyramid, 107
    Square pyramid, 108, 114
    Square tray, 103, 114
    Some new elliptical, 129-140
    Stadium, 115
    Standard normal, 192, 205
    Symmetric, Kotz-type, 129, 130
    T, 83, 190, 195, 196
    Totally Positive of Order Two, 162
    Type III, extreme value, 129
    Two-parameter Pareto, 153
    Uniform, 90, 103, 154
    Weibull, 3, 6, 7, 129, 130
    Wilcoxon, 39, 40

E

Econometrics,
    Statistical Aspects of, 20-28
Effect,
    Covariate, 117, 126
    Design, 98, 99
    Discrete randon, 182
    Genetic, 179-187
    Indirect, 126
    Of loci, 179
    QLT, 181
    Treatment, 36
    Weak, 56
Equation,
    Explicit, 123
    Matrix, 184
    Non-linear, 123
    Non-linear likelihood, 150
    Regression, 24, 85
    Renewal, 167, 168, 171, 177
EM Algorithm,
    E-step, 184
    For mixtures, 185
    M-step, 184
    Parameter estimation, 182-186
Environmental Issues, 1
Environment,
    Quality of, 13

Error,
- Asymptotic mean square, 154, 155
- Binomial, 117
- Coding, 76-77
- Estimated standard, 181
- Mean square, 154, 155
- Normal, 189
- Of the Hill estimator, 149
- Sampling, 46
- Square prediction, 191
- Standard, 186
- Standardized prediction, 196, 197
- Stochastic, 149
- Structure, 25
- Tukey's, standard, 33
- Two-dimensional, 162

Estimation,
- Basic, 4
- Conditional maximum likelihood, method, 144
- James-Stein, 32, 33
- Kernel density, 85-86
- Maximum Likelihood, 35, 182
- Method, for generalized latent variable model, 118, 120-123
- Minimax, 36
- Nonparametric, 91
- Of extreme-value index, 143, 149, 152, 153, 156
- Order-based, 5
- Over testing, 58
- Parameter, 143
- Parametric, theory, 142
- Robust, 153
- Semi-parametric, 149

Estimator,
- Based on Mean Excess Plots, 146-147
- Based on QQ plots, 146
- Best linear unbiased, 4
- Extreme value index, 141-159
- Generalized Jackknife, 149
- Hill, 144, 145, 146, 147, 148, 149, 151, 154, 155
- Hill, Smoothing, 151-152
- Kernel, 147-148
- 'k-records', 148
- Least squares, 189
- Moment, 144-145, 148, 149, 151, 155
- Moment-ratio, 145
- Moment, Smoothing, 152
- Peng's, 145
- Pickands, 143-144, 148, 149, 155
- Robust, based on Excess Plots, 152-153
- Semi-parametric, 150
- Smoothing, 151
- Smoothing and Robustifying, 150
- Theoretical comparison, 149
- W, 145

Experimental Design, 36, 97
- Efficient, 41

Experiments,
- Double-blinded controlled, 36
- Natural, 25, 66
- Of nature, 49
- Randomized, 24
- Randomized controlled, 66

Exponential,
- Case, 170, 171
- Distribution, 5, 81, 154, 170, 175
- Double, 10
- Family, 15, 16, 33, 120
- Mixtures, 81
- One-parameter, family, 15, 185

Extremes 2, 3, 5, 6
- Distributional behavior of, 5
- Upper, 3
- Value index, 141
- Value theory, 141

F

F distribution,
- Generalized form, 193

Factor
- General trend, 196
- Loadings, 87
- Scores, 16

Factorization, 15

Functions,
- Admissible decision, 23
- Bessel, 192
- Characteristic, 133-134, 136-137, 139
- Continuous, 204
- Cumulative distribution, 108
- Density, 88, 183, 193, 195
- Density, normal distribution, 84
- Distribution, 78, 141, 146
- Generalized Median Excess, 153
- Generating, 75, 77
- Hypergeometric, 131, 132, 136
- Kernel, 147
- Likelihood, 82, 182, 184
- Link, 118, 119
- Log-likelihood, 183

SUBJECT INDEX   221

Marginal density, resulting, 84
Mean excess, 146
Meijer's G, 136
Modified Bessel, 176
Monotonic, 176
n-variate distribution, 161
Power, 165
Probability density, 79, 81, 85, 103,
    192, 196, 201
Renewal, 167, 170
Response, approach, 117
Scale, 129
Single response, 120
Trimmed mean excess, 153

G

H

Hazards,
    Kaplan-Meier, Proportional, 32
Hypothesis,
    Consitutional, 62, 63

I

Independence,
    Conditional, 15, 25, 86, 87
    Local, 121
    Social, 125
Index
    Arbitrary positive, 207
    Extreme-value, estimators, 141-159
    Extreme-value, 141, 148, 149, 150,
        151, 152, 153, 154, 155, 156
    Poverty, 124
    Semi-parametric extreme-value,
        estimators, 143-144, 150
    Smoothing extreme-value,
        estimators, 151-152
    Tail, 141, 146, 153, 155
Indicators,
    Variable, 15
    Welfare, 124
Inference,
    Bayesian, 33
    Causal, 45
    Conditional, 43
    Ecological, 50
    Modes of, 22, 23, 24, 25
        Statistical, 1, 11, 22, 31, 44,
            61, 97
Interval,

Confidence, 36, 61, 62, 65,
    66, 91
Mapping model, 181, 186
Marker, 179
Testing, 181
Invariance, 35
    The role of, 53
Items,
    Polytomous, 119-120

J

Jacknife, 33, 43
    Algorithm, 149

K

L

Least squares, 30, 51
    Extended, 4
Life,
    Expectancy, 13
    Residual, 167, 174
    Testing applications, 81
Limits,
    Behavior of mean excess function,
        146
    Laws, 2, 3, 5
    Laws for maxima, 141
Likelihood
    Direct Interpretation of, 35
    Functions, modified, 22
    Log, 121, 122, 123
    Log, complete data, 183, 184
    Partial, 43
    Positive, ratio dependence, 164
LISCOMP, 18

M

Markov
    Chain, 73, 179
    Chain Monte Carlo, 22
    Models, 90
Mathematics,
    Academic, 34
    Of chance, 96
    Of Gauss, 52
Matrix,
    Augmented, 75

Diagonal, 75
Identity, 189
Low-order covariance, 56
P, 75
Sample covariance, 123
Transition probability, 73, 74
Variance/covariance, 4, 8, 9, 82, 87, 89, 185
Maximum
    Domain of attraction, 141
Maximum Likelihood, 182, 186
    Estimates, 179-187
    Estimation 31,33, 35, 56
    Theorem for mixtures, 92
MCMC, 32, 33, 92
Means,
    Adjusted, 186
    Conditional, of random component distributions, 118
    Of mixture models, 83
    Of graphical illustrations, 130
    Population, 18, 19
Measure,
    Departure from uniformity, 112-116
    Of dependence, 112
    Of location, 15, 16
    Of optimality, 154
    Product, 166
    Of reliability, 16, 17
    Summary, 14, 59
    Of uncertainty, 27, 65
Measurement Issues, 16-18, 20
    Problems, Social Sciences, 13-19
Metric,
    Distance, method of, 204, 208
    Ideal, 205, 206
    Ky-Fan, 205, 207
    Lévy-Prokhorov, 205, 206
    Minimal, 205
    Probability, 206, 207
    Scale, 117
    Variables, 117, 124
Metrical, 14, 15, 17
Memedian, 2, 7, 9, 10, 11 (7-11)
Methodology,
    Distribution-free outlier, 6
    Era, 32
    Monte-Carlo, 179
    Outlier, 5
    Of outliers, 5
Methods
    Bayesian statistical, 88

    Binomial and Poisson, 30
    Diagnostic, 154
    Empirical Bayes, 22
    Evaluation, 188
    Markov Chain Monte Carlo, 22
    Maximum Likelihood, 150
    Model-based, 5
    Monograph, 97
    Monte-Carlo, 149
    Multivariate extreme-value, 156
    Nonparametric, 32
    Nonparametric and Robust, 32
    Of ascertainment, 90
    Of block maxima, 142
    Of measurement, 101
    Of metric distances, 204, 208
    Of probability weighted moments, 150
    Of purification, 48
    Quantitative in econometrics, 20
    Regression, 51
    Representative, 97
    Selecting k, 153-154
    Semi-Parametric Estimation, 148-149
    Smoothing, 141
    Survey, 99, 101
    The Peaks Over Threshold Estimation, 149-150
Mixture,
    Finite, 83
    Gamma, 81
    Integral, 162
    k- finite step, 79
    Normal, 81
    Poisson, 81, 186
    Scale, 83
Models
    ANOVA, 84
    Bayes, 40
    Comparison problems, 188
    Competing, 188, 197
    Cox, 54
    Damage, 90
    Error-in variables, 85
    F_2-Generation, 181-182
    For probability sampling, 96-102
    Forecasting potential of, 188
    Generalized latent variable, 119
    Generalized linear, 4, 33, 36, 85, 117, 118-120, 126
    Generating, 90
    Genetic, 181
    Heteroscedastic, nonlinear, 185
    Hierarchical Bayes, 88
    Hierarchical generalized linear, 85

Inhomogeneity, 79-80
Kernel mixture, 86
Latent structure, 86-88
Latent variable, with covariates, 117-128
Linear, 189, 196
Linear forecasting, 188
Linear, rating the predictive ability, 189-190
Measurement error, 90
Mixed latent trait, 123
Mixture, 78-95
Multilevel, 89
One-factor, 125, 126
Other, 90-91
Overdispersion, 80
'Peaks Over Threshold', 149
Predictive, 188-202
Predictive adequacy of, 188, 190
Probabilistic, 21
Random coefficient regression, 85
Random effect, 85
Random effects, related, 84-85
Regression, 33, 45, 64, 65, 66, 189
Regression, in Epidemiology, 54-56
Regression, in Social Sciences, 51-53
Selection criteria, 124
Selecting, 124
Simple, 98
Specification, 55, 56
Statistical composite, interval mapping, 181
Statistical, Criteria for, 21
Two-factor, 125, 126
Variance components, 89
Crop-Yield, 196
Yule's, 53

N

Nonlinearity, 28
Nonparametrics, 36
Normality, 2
    Asymptotic, 143, 144, 145, 146, 148, 149, 151

O

Observations,
    Large, 151
    Multivariate, 4
Observational Studies,
    Interpretation of, 23-25
Odds ratio, 61, 62, 65
Optimality, 31, 154
Ordered Sample, 1, 8, 11
    Inference from, 2-5
Outliers, 1, 2, 3, 4, 5, 6, 11
    A Distribution-Free Approach to, 5-7
    Behavior, 5
    Discordant, 4
    Fundamental problem of, 5
    Identification, 3
    Possible new routes, 5-11
    Robustness studies, 83-84
Overdispersion, 78

P

Parameter,
    Additive, 181
    Arbitrary, 175
    Bandwidth, 147
    Canonical, 120
    Estimation by EM Algorithm, 182-186
    Extreme-value, 148, 150
    Genetic, 181
    Intercept, 85
    Kernel, 86
    Multiplicative spacing, 144
    Nuisance, 22
    Shape, 141, 142, 143
    Smoothing, 86
    Switching, 86
    Vector, 182
Physics,
    Maxwell, 96
    Social, 52
Plot
    Box and whisker, 2
    Hill, 151
    QQ, 146, 150, 156
    Mean, 156
    Mean excess, 146
    Pareto QQ, 146
Point,
    Extreme, 161, 162
Population,
    Actual, 98
    Combining samples, 99-100
    Complex, 98-99
    Experimental, 179
    Finite, 98

F_2, 179, 181, 186
Frame, 97, 98
IID, 99
Inhomogeneity, 80
Inhomogeneous, 79, 80, 81
Intercensal, estimates, 64
Mobile, 101
National, 97, 99
Normal, 83, 84
Of random individuals, 97
Reference, 63
Theoretical, of IID elements, 98

Principle,
    Plug-in, 36

Priors,
    Believable, 41
    Reference, 22
    Standardized reference, 22
    Subjective, 35

Probability,
    Posterior, 83

Problem,
    Bias, 151
    "big-data", 37
    Deconvolution, 90
    Industrial inspection, 23
    Non-Bayesian likelihood, 22
    Of comparative evaluation, 188, 189
    Of interpretation, 52
    Of pattern, 72

Process,
    Data collection, 21
    Data generation, 21
    Markov, 27
    Poisson, 175
    Renewal, 167, 168, 170, 171, 174, 176
    Renewal counting, 167
    Semi-Markov, 27

Q

QQ plots, 146, 150, 156

R

Random,
    Bivariate, variable, 104
    Coefficient regression model, 85
    Component distributions, 118
    Effects, 52
    Effects model, 84-85
    Error, 65
    Measurement error, 47
    n-dimensional vector, 129
    Number generators, 101
    Polynomials, 208
    Response variables, 118
    Risky variations, 100
    Sample, from a Cauchy distribution, 31
    Sample, 61
    Variate generation, 88-89

Randomization, 98

Rate,
    Convergence, 145, 203
    Convergence, estimates in functional limit theorems, 203-210
    False Discovery, 41

Reliability, 16-17, 26, 27

Regression,
    Approach, 154
    Coefficient, 23, 189
    Dependence, 162, 164-165
    Galton, 30
    Logistic, 32, 33
    Model formulation, 146
    Partial, coefficient, 24
    Partial, 181
    Poisson, 85
    Robust, 36
    Strongly positive, 165

Renewal,
    Counting process, 167
    Equations, 167, 168, 171, 176
    Functions, 167
    Paradox, 171-172
    Points, 167
    Process, 167, 168, 170, 171, 174, 176
    The lifespan of, 167-178

Replication, 98, 62, 64
    Bootstrap, 155

Residuals,
    Least absolute, 51
    Recursive, 190

Rules,
    Scoring, 188

S

Sample
    Averaging, periodic, 100
    Designs, 99

Finite, 5, 142
Of national populations, 99, 100
Order-based, 5
Ordering, 1
Periodic, 99
Population, combining, 99-100
Random, 121
Ranked set, 8, 9, 10, 11
Rolling, 99, 100

Sampling,
Chance, 101
Expectation, 100-101
Probability, 97, 98, 100, 101
Probability, new paradigms for, 96-102
Ranked set, 1, 5, 7, 9
Survey, 97, 98, 99, 100, 101
The paradigm of, 97-98

Science,
Information, 30

Sequence,
Of weights, 154
Optimal, 153, 155

Series,
Hypergeometric, 193
Single long, 27
Short, 27

Set,
Convex, 161
Ranked, BLUE, 10, 11
Ranked, sample, 8
Ranked, sample mean, 9, 10, 11
Ranked, sampling, 1, 5, 7, 9
Training, 82

Significance,
Fixed-level, 66
Observed level, 57
Statistical, 25, 53, 58
Substantive, 25
The search for, 56, 57, 58

Spaces,
Banach, 203

Statistic,
As a new paradigm, 96-97
Bayesian, 34, 36
Dixon, 4
Fisherian, 36
History of, 45-71
Order, 151
Pattern recognition, 82
Reduced order, 4
The student's t, 31
Upper-order, 154

Statistical,

Assumptions, 56
Century, 29-44
Models, 45
Optimality, 41
Philosophy, 34
Practice, 34
Thinking, 29

Stochastically,
Dominated, 172
Larger, 172

Sufficient, 15

System,
Multidimensional, 27

T

Test
F, 196
Of Discordancy, 3, 5
Permutation, 35
Statistic, 57
Wilcoxon's, 33, 39
Wilcoxon two-sample, 39

Testing,
Educational, 14
Goodness of Fit, 25-26
Statistical, 58

Theorem,
Bayes, 30, 88
Blackwell's, 167, 177
Central Limit, 2, 30, 204
Continuity, 203, 204
Dynkin-Lamperti, 173
Elementary renewal, 172
Fisher-Tippet's, 141
Functional limit, 203-210
Gauss-Markov, 4
Key Renewal, 167, 168, 169, 175
Kotlarski's, 192
Least Squares, 30
Yamukov's, 207

Theory,
Asymptotic, 22
Bayesian, 41
Coherent, 30
Distribution, 162
Era, 32
Extreme-value, 142, 143, 147, 149, 150
Fisher's optimality, 37
General renewal, 167, 172
Germ, 47

Time,
  Probability, 203
  Utility, 175
  Random, 175
Time series, 3, 22, 23, 25, 27, 124
Transform,
  Laplace, 175, 176, 177
  Mellin, 191, 192, 193
Treatment,
  Effect of, 46

U

Unbiasedness, 36

V

Value,
  Expected, 101
  Extreme, 17, 141
  Hill estimators, 151
  Genotypic, 181
  k-record, 148
  Mean, 18
  Missing, 182, 183
  Moderate, 164
  Ordered sample, 2
  P, 57, 59, 61, 62, 65, 66
  Potential ordered sample, 1
  Sample, 7
Variables,
  Binary response, 56
  Categorical, 14, 15, 17, 83
  Chi-square, 190
  Confounding, 45, 46
  Dependent, 191
  Dependent random, 189, 190, 193
  Discrete random, 165
  Distributed normal, 190
  Explanatory, 56, 117, 126, 181
  Explanatory, two observed, 124
  Gamma, distributed, 117
  Generic, 172
  Independent t, 190
  Intermediate, 58
  Latent, 14-16, 17, 18, 19, 118, 126
  Latent, the effect of, 117
  Manifest, 118, 120, 121
  Metric, 117, 124
  Mixed, manifest, 117, 118
  Normal, distributed, 123
  Normal error, 190
  Observed, 117, 118, 123, 126
  On welfare, 124-125
  Poisson, distributed, 123
  Random, 162, 164, 167, 170, 172, 208
  Random, Positive valued, 191
  Single latent, 124
  Totally positive of order two, 164
  Survey, 98, 99
Variance,
  Analysis of, 84
  Asymptotic, 146, 151, 152, 179, 185
  Components model, 89
  Covariance matrix, 4, 8, 82, 87, 89
  Diagonal covariance matrix, 9
  Formulas, 99
  Of the generic distribution, 171
  Posterior, 16
  Sample, 78
  Unit, 204, 208

W

X

Y

Z